Revise
AS

AQA
Biology

Jo   sett

Contents

Contents

Specification list

The specification labels on each page refer directly to the units in the exam specification, i.e. AQA ▶ 1.2 refers to unit 1, section 2.

AQA Biology

MODULE	SPECIFICATION TOPIC	CHAPTER REFERENCE	STUDIED IN CLASS	REVISED	PRACTICE QUESTIONS
	Disease and lifestyle	8.1, 8.2			
	Digestion	3.3			
	Biological molecules	1.1, 1.2, 1.3, 1.4			
	Enzymes	3.1, 3.2, 3.4			
Unit 1	Cells and membrane transport	2.1, 2.2, 2.3, 4.1, 4.2, 8.2			
	The lungs and lung disease	4.3, 8.1			
	The heart and heart disease	5.2, 8.1			
	Immunity	8.3, 8.4			
	Variation	7.2			
	DNA and the genetic code	6.1			
	Cell division	6.2			
	Biochemical differences	2.1, 1.2, 5.4			
	Cell differentiation	2.3			
Unit 2	Transport	5.1, 5.3, 5.5			
	Gaseous exchange	4.4, 5.5			
	Classification	7.1			
	Mutation and variation	7.2			
	Biodiversity	7.2			

Examination analysis

Unit 1 – Examination paper (100 UMS) (60 raw marks) 5–7 short answer questions plus 2 longer questions, a short comprehension and a short structured essay $1^1/_4$ hours 33.33% of the total AS marks 16.67% of the total A-level marks

Unit 2 – Examination paper (140 UMS) (85 raw marks) 7–9 short answer questions plus 2 longer questions (1 data handling and 1 assessing analysis and evaluation) $1^3/_4$ hours 46.67% of the total AS marks 23.33% of the total A-level marks

Unit 3 – Investigative and practical skills in AS Biology AS Centre Assessed Unit (60 UMS) (50 raw marks) Practical Skills Assessment (PSA) 6 marks, Investigative Skills Assignment (ISA) 44 marks, 20% of the total AS marks 10% of the total A-level marks

The AS/A2 Level Biology course

AS and A2

The AQA Biology A Level course being studied from September 2008 is in two parts, with a number of separate modules or units in each part. Most students will start by studying the AS (Advanced Subsidiary) course. Some will go on to study the second part of the A Level course, called A2. Advanced Subsidiary is assessed at the standard expected halfway through an A Level course, i.e. between GCSE and A Level. This means that the new AS and A2 courses are designed so that difficulty steadily increases:

- AS Biology builds from GCSE Science and Additional Science/Biology.
- A2 Biology builds from AS Biology.

How will you be tested?

Assessment units

AS Biology comprises three units or modules. The first two units are assessed by examinations. The third component involves practical assessment.

Centre-based practical assessment involves practical skills marked by your teacher. The marks can be adjusted by moderators appointed by the awarding body.

For AS Biology, you will be tested by three assessment units. For the full A Level in Biology, you will take a further three units. AS Biology forms 50% of the assessment weighting for the full A Level.

Tests are taken at two specific times of the year, January and June. It can be an advantage to you to take a unit test at the earlier optional time because you can re-sit the test. The best mark for each unit will be credited and the lower mark ignored.

Each unit can normally be taken in either January or June. Alternatively, you can study the whole course before taking any of the unit tests. There is a lot of flexibility about when exams can be taken and the following diagram shows just some of the ways that the assessment units may be taken for AS and A Level Biology.

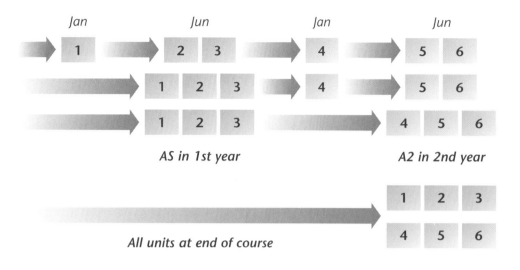

Jan   Jun   Jan   Jun

AS in 1st year   *A2 in 2nd year*

All units at end of course

A2 and synoptic assessment

Many students who have studied at AS Level may decide to go on to study A2. There are three further units or modules to be studied. The A Level specification includes a 'synoptic' assessment at the end of A2. Synoptic questions make use of concepts from earlier units, bringing them together in holistic contexts. Examiners will test your ability to inter-relate topics through the complete course from AS to A2.

Internal Assessment of Practical Skills

Internal assessment will form part of your A Level Biology course. It will be in the form of assessment of practical skills.

What skills will I need?

For AS Biology, you will be tested by assessment objectives: these are the skills and abilities that you should have acquired by studying the course. The assessment objectives for AS Biology are shown below.

Knowledge with understanding

- recall of facts, terminology and relationships
- understanding of principles and concepts
- drawing on existing knowledge to show understanding of the responsible use of biological applications in society
- selecting, organising and presenting information clearly and logically

Application of knowledge and understanding, analysis and evaluation

- explaining and interpreting principles and concepts
- interpreting and translating, from one to another, data presented as continuous prose or in tables, diagrams and graphs
- carrying out relevant calculations
- applying knowledge and understanding to familiar and unfamiliar situations
- assessing the validity of biological information, experiments, inferences and statements

You must also present arguments and ideas clearly and logically, using specialist vocabulary where appropriate. Remember to balance your argument!

Investigative and practical skills

One of the three AS units (and one of the A2 units) is designed to test your investigative and practical skills. Each unit is made up of two parts:
- **Practical Skills Assessment (PSA)**
- **Investigative Skills Assessment (ISA)**

The **PSA** involves your teacher marking you on a scale 0 to 6 based on your practical work throughout the course. They will be looking to see if you can...
- follow instructions
- select and use equipment
- carry out practical work in an organised and safe way.

The **ISA** carries 44 marks and is made up of three tasks:
- First you collect some practical data
- Then you process the data
- This is followed by a written paper relating to the experiment.

These ISA tasks are provided by AQA and you will carry them out under controlled conditions, being supervised by your teacher. Your teacher then marks them using a mark scheme provided by AQA. A task cannot be repeated but you may be given the opportunity to do more than one of each type of task. Your best mark on each will count. (There is an alternative to the ISA which is provided and marked by AQA. Only one of these is provided each year.)

It is important in these tasks to know the difference between **validity**, **reliability**, **precision** and **accuracy**.

- **Validity:** in a valid experiment you only investigate one variable at a time. All the other variables are kept constant. This means that you are really measuring what you are trying to measure.
- **Reliability:** an experiment is reliable if it can be repeated and similar results are obtained. Reliability is usually be increased by taking repeat readings and averaging them.
- **Accuracy:** an accurate measurement is one that is close to an accepted value. This accepted value is usually the work of many scientists and can be checked from accepted sources.
- **Precision:** If an experiment is precise it means that readings can be made to a number of significant digits and that similar results can be made each time.

The best way to describe the difference between accuracy and precision is to use the example of a darts board:

These darts are precise because they are all close to each other, but they are not very accurate.

These darts are precise because they are all close to each other and are accurate.

These darts are not very precise, nor very accurate.

Different types of questions in AS examinations

In AS Biology examinations different types of questions are used to assess your abilities and skills. Unit tests mainly use structured questions requiring both short-answers and more extended answers.

Short-answer questions

A question will normally begin with a brief amount of stimulus material. This may be in the form of a diagram, data or graph. A short-answer question may begin by testing recall. Usually this is followed up by questions which test understanding. Often you will be required to analyse data.

Short-answer questions normally have a space for your responses on the printed paper. The number of lines is a guide as to the amount of words you will need to answer the question. The number of marks indicated on the right side of the paper shows the number of marks you can score for each question part.

Here are some examples. (The answers are shown in blue)

The diagram shows part of a DNA molecule.

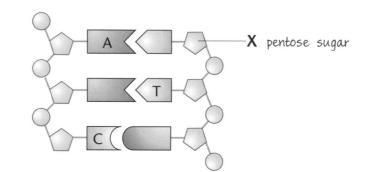

(a) Label part X. [1]

(b) Complete the diagram by writing a letter for each missing organic base in each empty box. [1]

(b) How do two strands of DNA join to each other?

The organic bases ✓ link the strands by hydrogen bonds ✓ [2]

Structured questions

Structured questions are in several parts. The parts are usually about a common context and they often progress in difficulty as you work through each of the parts. They may start with simple recall, then test understanding of a familiar or unfamiliar situation. If the context seems unfamiliar the material will still be centred around concepts and skills from the Biology specification. (If a student can answer questions about unfamiliar situations then they display understanding rather than simple recall.)

The most difficult part of a structured question is usually at the end. Ascending in difficulty, a question allows a candidate to build in confidence. Right at the end technological and social applications of biological principles give a more demanding challenge. Most of the questions in this book are structured questions. This is the main type of question used in the assessment of AS Biology.

When answering structured questions, do not feel that you have to complete a question before starting the next. Answering a part that you are sure of will build your confidence. If you run out of ideas go on to the next question. This will be more profitable than staying with a very difficult question which slows down progress. Return at the end when you have more time.

Here is an example of a structured question which becomes progressively more demanding.

Question

The diagram shows the molecules of a cell surface membrane.

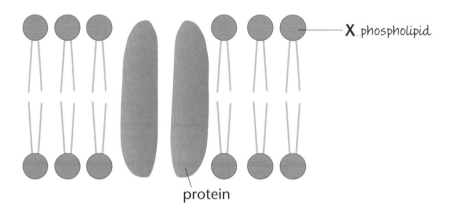

X phospholipid

protein

(a) (i) Label molecule X. [1]

(ii) The part of molecule X facing the outside of a cell is hydrophilic. What does this mean?

water loving/water attracting [1]

(iii) Describe **one** feature of the part of molecule X which faces inwards.

Hydrophobic/ water hating fatty acid residues [1]

(b) Explain how the protein shown in the diagram can actively transport the glucose molecule into the cell.

Energy is released from mitochondria near the channel protein the channel protein opens [3]

Note the help given in diagrams. The labelling of the protein molecule may trigger the memory so that the candidate has to make a small step to link the 'channel' function to this diagram. Examiners give clues! Expect more clues at AS Level than at A2 Level.

Extended answers

In AS Biology, questions requiring more extended answers will usually form part of structured questions. They will normally appear at the end of a structured question and will typically have a value of three to six marks. Longer answers are allocated more lines, so you can use this as a guide as to the extent of your answer. The mark allocation is a guide as to how many points you need to make in your response. Often for an answer worth six marks the mark scheme could have eight creditable answers. You are awarded up to the maximum, six in this instance.

Depending on the awarding body, longer extended questions may be set. These are often open response questions. These questions may be worth up to ten marks for full credit. Extended answers are used to allocate marks for the **quality of written communication**.

Candidates are assessed on their ability to use a suitable style of writing, and organise relevant material, both logically and clearly. The use of specialist biological words in context is also assessed. Spelling, punctuation and grammar are also taken into consideration. Here is a longer extended response question.

Question

Give an account of the effects of sewage entry into a river and explain the possible consequences to organisms downstream.

The sewage enters the river and is decomposed by bacteria. ✓ These bacteria are saprobiotic ✓ they produce nitrates which act as a fertiliser. ✓ Algae form a blanket on the surface ✓ light cannot reach plants under the algae so these plants die. ✓ Bacteria decompose the dead plants ✓ the bacteria use oxygen/bacteria are aerobic ✓ fish die due to lack of oxygen ✓ Tubifex worms or bloodworms increase near sewage entry ✓ mayfly larvae cannot live close to sewage entry/mayfly larvae appear a distance downstream where oxygen levels return. ✓

10 marking points → [7]

Remember that mark schemes for extended questions often exceed the question total, but you can only be awarded credit up to a maximum. Examiners sometimes build in a hurdle, e.g. in the above responses, references to one organism which increases in population is worthy of a mark, and another which decreases in population is worth another. Continually referring to different species which repeat a growth pattern will not gain further credit.

Exam technique

AS Biology builds from grade C in GCSE Science and GCSE Additional Science (combined) or GCSE Biology. This study guide has been written so that you will be able to tackle AS Biology from a GCSE science background.

You should not need to search for important Biology from GCSE science because this has been included where needed in each chapter. If you have not studied science for some time, you should still be able to learn AS Biology using this text alone.

What are examiners looking for?

Whatever type of question you are answering, it is important to respond in a suitable way. Examiners use instructions to help you to decide the length and depth of your answer. The most common words used are given below, together with a brief description of what each word is asking for.

Define

This requires a formal statement. Some definitions are easy to recall.

Define the term active transport.

This is the movement of molecules from where they are in lower concentration to where they are in higher concentration. The process requires energy.

Other definitions are more complex. Where you have problems it is helpful to give an example.

Define the term endemic.

This means that a disease is found regularly in a group of people, district or country. Use of an example clarifies the meaning. Indicating that malaria is invariably found everywhere in a country, confirms understanding.

Explain

This requires a reason. The amount of detail needed is shown by the number of marks allocated.

Explain the difference between resolution and magnification.

Resolution is the ability to be able to distinguish between two points whereas magnification is the number of times an image is bigger than an object itself.

State

This requires a brief answer without any reason.

State one role of blood plasma in a mammal.

Transport of hormones to their target organs.

List

This requires a sequence of points with no explanation.

List the abiotic factors which can affect the rate of photosynthesis in pond weed.

carbon dioxide concentration; amount of light; temperature; pH of water

Describe

This requires a piece of prose which gives key points. Diagrams should be used where possible.

> Describe the nervous control of heart rate.

> The medulla oblongata ✓ of the brain connects to the sino atrial node in the right atrium, wall ✓ via the vagus nerve and the sympathetic nerve ✓ the sympathetic nerve speeds up the rate ✓ the vagus nerve slows it down. ✓

Discuss

This requires points both for and against, together with a criticism of each point. (**Compare** is a similar command word.)

> Discuss the advantages and disadvantages of using systemic insecticides in agriculture.

> Advantages are that the insecticides kill the pests which reduce yield ✓ they enter the sap of the plants so insects which consume sap die ✓ the insecticide lasts longer than a contact insecticide, 2 weeks is not uncommon ✓

> Disadvantages are that insecticide may remain in the product and harm a consumer e.g. humans ✓ it may destroy organisms other than the target ✓ no insecticide is 100% effective and develops resistant pests. ✓

Suggest

This means that there is no single correct answer. Often you are given an unfamiliar situation to analyse. The examiners hope for logical deductions from the data given and that, usually, you apply your knowledge of biological concepts and principles.

> The graph shows that the population of lynx decreased in 1980. Suggest reasons for this.

> Weather conditions prevented plant growth ✓ so the snowshoe hares could not get enough food and their population remained low ✓ so the lynx did not have enough hares(prey) to predate upon. ✓ The lynx could have had a disease which reduced numbers. ✓

Calculate

This requires that you work out a numerical answer. Remember to give the units and to show your working, marks are usually available for a partially correct answer. If you work everything out in stages write down the sequence. Otherwise if you merely give the answer and it is wrong, then the working marks are not available to you.

> Calculate the Rf value of spot X. (X is 25 mm from start and solvent front is 100 mm)

$$Rf = \frac{\text{distance moved by spot}}{\text{distance moved by the solvent front}}$$

$$= \frac{25 \text{ mm}}{100 \text{ mm}}$$

$$= 0.25$$

Outline

This requires that you give only the main points. The marks allocated will guide you on the number of points which you need to make.

Outline the use of restriction endonuclease in genetic engineering.

The enzyme is used to cut the DNA of the donor cell. ✓

It cuts the DNA up like this A T|G C C G A T = A T   + G C C G A T ✓
                                 T A C G G C|T A   T A C G G C   T A

The DNA in a bacterial plasmid is cut with the same restriction endonuclease. ✓

The donor DNA will fit onto the sticky ends of the broken plasmid. ✓

If a question does not seem to make sense, you may have mis-read it. Read it again!

Some dos and don'ts

Dos

Do answer the question
No credit can be given for good Biology that is irrelevant to the question.

Do use the mark allocation to guide how much you write
Two marks are awarded for two valid points – writing more will rarely gain more credit and could mean wasted time or even contradicting earlier valid points.

Do use diagrams, equations and tables in your responses
Even in 'essay style' questions, these offer an excellent way of communicating biology.

Do write legibly
An examiner cannot give marks if the answer cannot be read.

Do write using correct spelling and grammar. Structure longer essays carefully
Marks are now awarded for the quality of your language in exams.

Don'ts

Don't fill up any blank space on a paper
In structured questions, the number of dotted lines should guide the length of your answer.

If you write too much, you waste time and may not finish the exam paper. You also risk contradicting yourself.

Don't write out the question again
This wastes time. The marks are for the answer!

Don't contradict yourself
The examiner cannot be expected to choose which answer is intended. You could lose a hard-earned mark.

Don't spend too much time on a part that you find difficult
You may not have enough time to complete the exam. You can always return to a difficult calculation if you have time at the end of the exam.

What grade do you want?

Everyone would like to improve their grades but you will only manage this with a lot of hard work and determination. You should have a fair idea of your natural ability and likely grade in biology and the hints below offer advice on improving that grade.

For a Grade A

You will need to be a very good all-rounder.

- You must go into every exam knowing the work extremely well.
- You must be able to apply your knowledge to new, unfamiliar situations.
- You need to have practised many, many exam questions so that you are ready for the type of question that will appear.

The exams test all areas of the syllabus and any weaknesses in your biology will be found out. There must be no holes in your knowledge and understanding. For a Grade A, you must be competent in all areas.

For a Grade C

You must have a reasonable grasp of biology but you may have weaknesses in several areas and you will be unsure of some of the reasons for the biology.

- Many Grade C candidates are just as good at answering questions as the Grade A students but holes and weaknesses often show up in just some topics.
- To improve, you will need to master your weaknesses and you must prepare thoroughly for the exam. You must become a better all-rounder.

For a Grade E

You cannot afford to miss the easy marks. Even if you find biology difficult to understand and would be happy with a Grade E, there are plenty of questions in which you can gain marks.

- You must memorise all definitions.
- You must practise exam questions to give yourself confidence that you do know some biology. In exams, answer the parts of questions that you know first. You must not waste time on the difficult parts. You can always go back to these later.
- The areas of biology that you find most difficult are going to be hard to score on in exams. Even in the difficult questions, there are still marks to be gained. Show your working in calculations because credit is given for a sound method. You can always gain some marks if you get part of the way towards the solution.

What marks do you need?

The table below shows how your average mark is transferred into a grade.

average	80%	70%	60%	50%	40%
grade	A	B	C	D	E

To achieve an A* grade, you need to achieve a...

- grade A overall (80% or more on uniform mark scale) for the **whole** A level qualification
- grade A* (90% or more on the uniform mark scale) across your A2 units.

A* grades are awarded for the A level qualification only and not for the AS qualification or individual units.

Four steps to successful revision

Step 1: Understand

- Study the topic to be learned slowly. Make sure you understand the logic or important concepts.
- Mark up the text if necessary – underline, highlight and make notes.
- Re-read each paragraph slowly.

GO TO STEP 2

Step 2: Summarise

- Now make your own revision note summary:
  What is the main idea, theme or concept to be learnt?
  What are the main points? How does the logic develop?
  Ask questions: Why? How? What next?
- Use bullet points, mind maps, patterned notes.
- Link ideas with mnemonics, mind maps, crazy stories.
- Note the title and date of the revision notes
  (e.g. Biology: Cells, 3rd March).
- Organise your notes carefully and keep them in a file.

This is now in **short-term memory**. You will forget 80% of it if you do not go to Step 3.
GO TO STEP 3, but first take a 10 minute break.

Step 3: Memorise

- Take 25 minute learning 'bites' with 5 minute breaks.
- After each 5 minute break test yourself:
  Cover the original revision note summary.
  Write down the main points.
  Speak out loud (record on tape).
  Tell someone else.
  Repeat many times.

The material is well on its way to **long-term memory**.
You will forget 40% if you do not do step 4. **GO TO STEP 4**

Step 4: Track/Review

- Create a Revision Diary (one A4 page per day).
- Make a revision plan for the topic, e.g. 1 day later, 1 week later, 1 month later.
- Record your revision in your Revision Diary, e.g.
  Biology: Cells, 3rd March 25 minutes
  Biology: Cells, 5th March 15 minutes
  Biology: Cells, 3rd April 15 minutes
  ... and then at monthly intervals.

Chapter 1
Biological molecules

The following topics are covered in this chapter:

- Carbohydrates
- Lipids
- Proteins

1.1 Carbohydrates

After studying this section you should be able to:

- recall the main elements found in carbohydrates
- recall the structure of glucose, fructose, lactose, sucrose, starch, glycogen and cellulose
- recall the role of glucose, starch, glycogen, cellulose and pectin

LEARNING SUMMARY

Structure of carbohydrates

AQA   1.2, 2.4

Monosaccharides

All carbohydrates are formed from the elements carbon (C), hydrogen (H) and oxygen (O). The formula of a carbohydrate is always $(CH_2O)_n$. The n represents the number of times the basic CH_2O unit is repeated, e.g. where n = 6 the molecular formula is $C_6H_{12}O_6$. This is the formula shared by glucose and other simple sugars like fructose. These simple sugars are made up from a single sugar unit and are known as **monosaccharides**.

The molecular formula, $C_6H_{12}O_6$, does not indicate how the atoms bond together. Bonded to the carbon atoms are a number of – H and – OH groups. Different positions of these groups on the carbon chain are responsible for different properties of the molecules. The structural formulae of α and β glucose are shown below.

> These molecules are mirror images of each other. When molecules have the same molecular formula but different structural formulae, they are known as **isomers**. Isomers have different properties to each other.

α glucose   β glucose

Glucose is so small that it can pass through the villi and capillaries into our bloodstream. The molecules subsequently release energy as a result of respiration. Simple glucose molecules are capable of so much more than just releasing energy. They can combine with others to form bigger molecules.

Disaccharides

Each glucose unit is known as a **monomer** and is capable of linking others. This diagram shows two molecules of α glucose forming a **disaccharide**.

> In your examinations look for different monosaccharides being given, like fructose or β glucose. You may be asked to show how they bond together. The principle will be exactly the same.

α glucose   α glucose

H_2O condensation reaction

maltose

A **condensation** reaction means that as two carbohydrate molecules bond together a water molecule is produced. The link formed between the two glucose molecules is known as a **glycosidic bond**.

A glycosidic bond can also be broken down to release separate monomer units. This is the opposite of the reaction shown above. Instead of water being given off, a water molecule is needed to break each glycosidic bond. This is called **hydrolysis** because water is needed to split up the bigger molecule.

<table>
<tr><td>'**Lysis**' literally means '**splitting**'. In hydrolysis water is needed in the reaction to break down the molecule.</td></tr>
</table>

Different disaccharides are made by joining together different monosaccharides.

Disaccharide	Component monosaccharides
lactose	glucose + galactose
maltose	glucose + glucose
sucrose	glucose + fructose

Polysaccharides

Like disaccharides, **polysaccharides** consist of monomer units linked by the glycosidic bond. However, instead of just two monomer units they can have many. Chains of these 'sugar' units are known as **polymers**. These larger molecules have important structural and storage roles.

Starch is a polymer of the sugar, α glucose. The diagram below shows part of a starch molecule.

Notice the five glycosidic bonds on just a small part of a starch molecule.

part of a branched section of a starch molecule

This type of starch molecule is called **amylopectin** and it has a branched structure. Starch also contains **amylose**. This does not contain branches but the chain of glucose units forms a helix. **Glycogen** is similar in structure to amylopectin but with more branches. **Cellulose** is also a polymer of glucose units, but this time the units are β glucose.

How useful are polysaccharides?

- **Starch** is stored in organisms as a future energy source, e.g. potato has a high starch content to supply energy for the buds to grow at a later stage.
- **Glycogen** is stored in the liver, which releases glucose for energy in times of low blood sugar.

Both starch and glycogen are insoluble which enables them to remain inside cells.

The many branches in the amylopectin molecule means that enzymes can digest the molecule rapidly.

- **Cellulose** has long molecules which help form a tough protective layer around plant cells, the cell wall. Each cellulose molecule has up to 10 000 β glucose units. Each molecule can form cross-links with other cellulose molecules forming fibres. This makes cellulose fibres very strong.

Pectins help cells to bind together.

- **Pectins** are used alongside cellulose in the cell wall. They are polysaccharides which are bound together by calcium pectate.

Together the cellulose and pectins give exceptional mechanical strength. The cell wall is also permeable to a wide range of substances.

1.2 Lipids

After studying this section you should be able to:

- *recall the main elements found in lipids*
- *recall the structure of lipids*
- *distinguish between saturated and unsaturated fats*
- *recall the role of lipids*

LEARNING SUMMARY

What are lipids?

AQA 1.3

Lipids include **oils, fats** and **waxes**. They consist of exactly the same elements as carbohydrates, i.e. carbon (C), hydrogen (H) and oxygen (O), but their proportion is different. There is always a high proportion of carbon and hydrogen, with a small proportion of oxygen.

The diagram below shows the structural formula of the most common type of lipid called a **triglyceride**.

R represents groups such as CH_3 or C_2H_5

a **triglyceride** fat

Triglycerides are formed when fatty acids react with glycerol. During this reaction water is produced, a further example of a condensation reaction. The essential bond is the **ester bond**.

Note that water is produced during triglyceride formation. This is another example of a condensation reaction. Different triglyceride fats are formed from different fatty acids.

3 fatty acids   glycerol   a triglyceride fat   water

Triglycerides can be changed back into the original fatty acids and glycerol. Enzymes are needed for this transformation together with water molecules. Remember, an enzyme reaction which requires water to break up a molecule is known as **hydrolysis**.

The hydrocarbon chains are so long that they are often represented by the acid group (–COOH) and a zig-zag line.

unsaturated
∿∿∿=∿∿COOH

saturated
∿∿∿∿∿COOH

Saturated fats lead to cholesterol build up in the blood vessels so many people try to avoid them in their diet. Blocked vessels cause heart attacks.

Note that when saturated fatty acids react with glycerol they make saturated fats. Conversely, unsaturated fatty acids make unsaturated fats.

Never give an ambiguous answer like this: 'fats keep an animal warm'. The examiner does not know whether you are referring to energy release by mitochondria or the insulation property. Be specific, otherwise you will not be credited.

What are saturated and unsaturated fats?

The answer lies in the types of fatty acid used to produce them.

a **saturated** fatty acid

an **unsaturated** fatty acid

> **KEY POINT**
> Saturated fatty acids have no C=C (double bonds) in their hydrocarbon chain, but unsaturated fatty acids do. This is the difference.

How useful are lipids?

Like carbohydrates, lipids are used as an **energy** supply, but a given amount of lipid releases more energy than the same amount of carbohydrate. Due to their **insolubility** in water and **compact** structure, lipids have long-term **storage** qualities. Adipose cells beneath our skin contain large quantities of fat which **insulate** us and help to maintain body temperature. Fat gives **mechanical** support around our soft organs and even gives **electrical insulation** around our nerve axons.

An aquatic organism such as a dolphin has a large fat layer which:

- is an energy store
- is a thermal insulator
- helps the animal remain buoyant.

The most important role of lipids is their function in cell membranes. To fulfil these functions a triglyceride fat is first converted into a **phospholipid**.

triglyceride   phosphoric acid   phospholipid

Phosphoric acid replaces one of the fatty acids of the triglyceride. The new molecule, the phospholipid, is a major component of cell membranes. Cell membranes also contain the lipid cholesterol. The following diagram represents a phospholipid.

This is the polar head.
It attracts water.
This means it is **hydrophilic**.

Remember this simplified diagram when you learn about the cell membrane.

These are fatty acid tails.
They repel water.
This means they are **hydrophobic**.

a phospholipid

1.3 Proteins

After studying this section you should be able to:

- *recall the main elements found in proteins*
- *recall how proteins are constructed*
- *recall the structure of proteins*
- *recall the major functions of proteins*

LEARNING SUMMARY

The building blocks of proteins

AQA 1.2

Like carbohydrates and lipids, proteins contain the elements carbon (C), hydrogen (H) and oxygen (O), but in addition they **always** contain **nitrogen** (N). Sulfur is also often present.

Before understanding how proteins are constructed, the structure of **amino acids** should be noted. The diagram below shows the general structure of an amino acid.

Just like the earlier carbohydrate and lipid molecules, '**R**' represents groups such as $-CH_3$ and $-C_2H_5$. There are about 20 commonly found amino acids but you will not need to know them all. Instead, learn the basic structure shown opposite.

an amino acid

How is a protein constructed?

The process begins by amino acids bonding together. The diagram shows two amino acids being joined together by a **peptide bond**.

This is another example of a condensation reaction as water is produced as the dipeptide molecule is assembled.

Note that the peptide bonds can be broken down by a hydrolysis reaction.

When many amino acids join together a long-chain **polypeptide** is produced. The linking of amino acids in this way takes place during protein synthesis (see page 74). There are around 20 different amino acids. Organisms join amino acids in different linear sequences to form a variety of polypeptides, then build these polypeptides into complex molecules, the **proteins**. Humans need eight essential amino acids as adults and ten as children, all the others can be made inside the cells.

The sequence of amino acids along a polypeptide is controlled by another complex molecule, DNA (see the genetic code, page 73).

Levels of organisation in proteins

AQA 1.2

Primary protein structure

This is the **linear sequence** of amino acids.

Secondary protein structure

Polypeptides become twisted or coiled. These shapes are known as the **secondary structure**. There are two common secondary structures; the α-helix and the β-pleated sheet.

Both secondary structures give additional strength to proteins. The α-helix helps make tough fibres like the protein in your nails, e.g. keratin. The β-pleated sheet helps make the strength-giving protein in silk, fibroin. Many proteins are made from both α-helix and β-pleated sheet.

The polypeptides are held in position by hydrogen bonds. In both α-helices and β-pleated sheets the **C = O** of one amino acid bonds to the **H–N** of an adjacent amino acid like this: C = O --- H–N.

$$----- = \text{hydrogen bonds}$$

An α-helix is a tight, twisted strand; a β-pleated sheet is where a zig-zag line of amino acids bonds with the next, and so on. This forms a sheet or ribbon shape.

coiled α-helix structure   **a fibrous protein**

The protein shown only achieves a secondary structure as the simple α-helix polypeptides do not undergo further folding.

This is the structure of a fibrous protein. It is made of three α-helix polypeptides twisted together.

Tertiary protein structure

Note that the specific contours of proteins have extremely significant roles in life processes. (See enzymes page 40.)

This is when a polypeptide is **folded** into a **precise** shape. The polypeptide is held in 'bends' and 'tucks' in a **permanent** shape by a range of bonds including:

* **disulfide** bridges (sulfur–sulfur bonds)
* hydrogen bonds
* ionic bonds
* hydrophobic and hydrophilic interactions.

Some proteins are folded up into a spherical shape. They are called **globular proteins**. They are soluble in water. Other proteins, called **fibrous proteins**, form long chains. They are insoluble and usually perform structural functions.

This is the structure of a **globular** protein. It is made of an α-helix and a β-pleated sheet. Precise shapes are formed with specific contours.

Note that some proteins do not have a quaternary structure. If they consist of just one folded polypeptide then they are classified as having tertiary structure. If they are simple fibres of α-helices or β-pleated sheets then they have only secondary protein structure.

Quaternary protein structure

Some proteins consist of **different polypeptides** bonded together to form extremely intricate shapes. A haemoglobin molecule is formed from four separate polypeptide chains. It also has a haem group, which contains iron. This inorganic group is known as a **prosthetic group** and in this instance aids oxygen transport.

α-helix — — β-pleated sheet

— disulfide bridge

a quaternary structure

How useful are proteins?

AQA 1.2

Proteins can be just as beneficial as carbohydrates and lipids in releasing energy. Broken down into their component amino acids, they liberate energy during respiration. The list below shows **important** uses of proteins:

- **cell-membrane proteins** transport substances across the membrane for processes such as facilitated diffusion and active transport
- **enzymes** catalyse biochemical reactions, e.g. pepsin breaks down protein into polypeptides
- **hormones** are passed through the blood and trigger reactions in other parts of the body, e.g. insulin regulates blood sugar
- **immuno-proteins**, e.g. antibodies are made by lymphocytes and act against antigenic sites on micro-organisms
- **structural proteins** give strength to organs, e.g. collagen makes tendons tough
- **transport proteins**, e.g. haemoglobin transports oxygen in the blood
- **contractile proteins**, e.g. actin and myosin help muscles shorten during contraction
- **storage proteins**, e.g. aleurone in seeds helps germination, and casein in milk helps supply valuable protein to babies
- **buffer proteins**, e.g. blood proteins, due to their charge, help maintain the pH of plasma.

Progress check

1 List the sequence of structures in a globular protein such as haemoglobin.

2 The following statements refer to proteins used for different functions in the body. The list gives the name of different types of protein. Match the name of each type of protein with the correct statement.

(i) transport proteins
(ii) immuno-proteins
(iii) storage proteins
(iv) buffer proteins
(v) cell-membrane proteins

(vi) contractile proteins
(vii) enzymes
(viii) structural proteins
(ix) hormones

A haemoglobin is used to transport oxygen in blood.

B aleurone in seeds is a source of amino acids as it is broken down during germination.

C actin and myosin help muscles shorten during contraction.

Progress check

D antibodies made by lymphocytes against antigens.

E blood proteins, due to their charge, help maintain the pH of plasma.

F used to transport substances across the membrane for processes such as facilitated diffusion.

G passed through blood, used to trigger reactions in other parts of the body, e.g. FSH stimulates a primary follicle.

H used to catalyse biochemical reactions, e.g. amylase breaks down starch into maltose.

I used to give strength to organs, e.g. collagen makes tendons tough.

2 A (i), B (iii), C (vi), D (ii), E (iv), F (v), G (ix), H (vii), I (viii).
1 primary structure: amino acids linked in a linear sequence; secondary structure: α-helix or β-pleated sheet; tertiary structure: further folding of polypeptide held by disulfide bridges, ionic bonds, and hydrogen bonds; quaternary structure: two or more polypeptides bonded together.

1.4 Biochemical tests

After studying this section you should be able to:

● *describe biochemical tests for carbohydrates, lipids and proteins*

L E A R N I N G S U M M A R Y

Biochemical tests

AQA 1.2

All the biochemical tests need to be learned. This work is good value because they are regularly tested in 2 or 3 mark question components.

Tests for carbohydrates in the laboratory

Benedict's test used to identify reducing sugars (monosaccharides and some disaccharides)
- Add Benedict's solution to the chemical sample and heat.
- The solution changes from blue to brick-red or yellow if a reducing sugar is present.

Non-reducing sugar test used to test for non-reducing sugars, e.g. the disaccharide, sucrose
- First a Benedict's test is performed.
- If the Benedict's test is negative, the sample is hydrolysed by heating with hydrochloric acid, then neutralised with sodium hydrogen carbonate.
- This breaks the glycosidic bond of the disaccharide, releasing the monomers.
- A second Benedict's test is performed which will be positive because the monomers are now free.

Starch test
- Add iodine solution to the sample.
- If starch is present the colour changes to blue-black.

Tests for lipids in the laboratory

Emulsion test used to identify fats and oils
- Add ethanol to the sample, shake, then pour the mixture into water.
- If fats or oils are present then a white emulsion appears at the surface.

Tests for proteins in the laboratory

Biuret test used to identify any protein
- Add dilute sodium hydroxide and dilute copper sulfate to the sample.
- A violet colour appears if a protein is present.

Quantitative tests

The above tests can be used to see if the different chemicals are present or absent. They are called **qualitative tests**.

They do not provide any information about how much of the chemical is present unless they are modified. One way to do this is to measure the intensity of the colour produced. This can be done with a machine called a **colorimeter**. It shines a light through the solution and measures how much light gets through. The lighter the colour, the more light passes through and so the less chemical is present. This is a **quantitative test**.

Sample questions and model answers

1 Below are the structures of two glucose molecules.

(a) Complete the equation to show how the molecules react to form a glycosidic bond and the molecule produced.

Remember that you will be given molecule structures. These stimulate your memory which helps you work out the answer.

(b) Which form of glucose molecules is shown? Give a reason for your answer. [2]
α glucose, because the −OH groups on carbon atom 1 are down

(c) State the type of reaction which takes place when the two molecules shown above react together. [2]
Condensation.

The correct answer here is condensation. A regular error in questions like this is to give the wrong reaction, i.e. hydrolysis. Revise carefully then you will make the correct choice.

2 The diagram below shows a globular protein consisting of four polypeptide chains.

Look out for similar structures in your examinations. The proteins given may be different, but the principles remain the same.

(a) Use your own knowledge and the information given to explain how this protein shows primary, secondary, tertiary and quaternary structure. [5]
Primary structure: it is formed from chains of amino acids; it has polypeptides made of a linear sequence of amino acids.
Secondary structure: it has an α-helix, it has a β-pleated sheet.
Tertiary structure: the polypeptides are folded, the folds are held in position by disulfide bridges.
Quaternary structure: there are four polypeptides in this protein. Two or more are bonded together to give a quaternary structure.

(b) Name and describe a test which would show that haemoglobin is a protein. [3]
The Biuret test.
Take a sample of haemoglobin and add water, sodium hydroxide and copper sulfate.
The colour of the mixture shows as violet or mauve if the sample is a protein.

Most examinations include at least one biochemical test.

Practice examination questions

Try all of the questions and check your answers with the mark scheme on page 99

1 (a) Complete the equation below to show the breakdown of a triglyceride fat into fatty acids and glycerol. [2]

a triglyceride fat   3 molecules of water

(b) Describe a biochemical test which would show if a sample contained fat. [3]

2 The diagram below shows a polypeptide consisting of 15 amino acids.

amino acid

(group X)

(a) Name the bond between each pair of amino acids in this polypeptide. [1]

(b) What is group X? [1]

(c) Which level of protein structure is shown by this polypeptide? Give a reason for your answer. [2]

Chapter 2
Cells

The following topics are covered in this chapter:

- *The ultra-structure of cells*
- *Isolation of cell organelles*

- *Specialisation of cells*

2.1 The ultra-structure of cells

After studying this section you should be able to:

- identify cell organelles and understand their roles
- recall the differences between prokaryotic and eukaryotic cells

LEARNING SUMMARY

Cell organelles

AQA ▶ 1.3, 2.4

The cell is the basic functioning unit of organisms in which chemical reactions take place. These reactions involve energy release needed to support life and build structures. Organisms consist of one or more cells. The amoeba is composed of one cell, whereas millions of cells make up a human.

> Every cell possesses internal coded instructions to control cell activities and development. Cells also have the ability to continue life by some form of cell division.
>
> **KEY POINT**

Organelles are best seen with the aid of an electron microscope.

The ultra-structure of a cell can be seen using an electron microscope. Sub-cellular units called **organelles** become visible. Each organelle has been researched to help us understand more about the processes of life.

The animal cell and its organelles

The diagram below shows the organelles found in a typical animal cell.

A plant cell has all of the same structures plus:
- a cellulose cell wall
- chloroplasts (some cells)
- a sap vacuole with tonoplast.

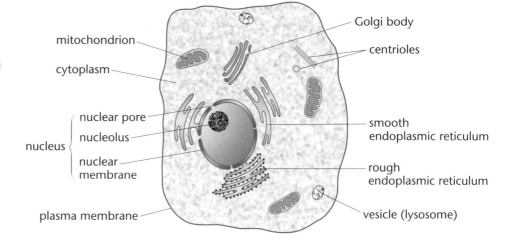

mitochondrion

cytoplasm

nucleus { nuclear pore, nucleolus, nuclear membrane

plasma membrane

Golgi body

centrioles

smooth endoplasmic reticulum

rough endoplasmic reticulum

vesicle (lysosome)

Cell surface (plasma) membrane

This covers the outside of a cell and consists of a **double** layered sheet of phospholipid molecules interspersed with proteins. It separates the cell from the outside environment, gives physical protection and allows the import and export of **selected** chemicals.

Plasma membrane

Nucleus

This controls all cellular activity using coded instructions located in DNA. These coded instructions enable the cell to make specific proteins. RNA is produced in the nucleus and leaves via the nuclear pores. The nucleus stores, replicates and decodes DNA.

nucleus   mitochondrion

Be ready to identify all cell organelles in either a diagram or electron micrograph. A mitochondrion is often sausage shaped but the end view is circular. Look out for the internal membranes.

Mitochondria

These consist of a double membrane enclosing a semi-fluid matrix. Throughout the matrix is an internal membrane, folded into cristae. The cristae and matrix contain enzymes which enable this organelle to carry out aerobic respiration. It is the key organelle in the release of energy, making ATP available to the cell.

Mitochondria are needed for many energy requiring processes in the cell, including active transport and the movement of cilia.

Cytoplasm

Cytoplasm is often seen as grey and granular. If the image is 'clear' then you are probably looking at a vacuole.

Each organelle in a cell is suspended in a semi-liquid medium, the cytoplasm. Many ions are dissolved in it. It is the site of many chemical reactions.

Ribosomes

Look for tiny dots in the cytoplasm. They will almost certainly be ribosomes. A membrane adjacent to a line of ribosomes is probably the rough endoplasmic reticulum.

There are numerous **ribosomes** in a cell, located along **rough endoplasmic reticulum**. They aid the manufacture of proteins, being the site where mRNA meets tRNA so that amino acids are bonded together.

Endoplasmic reticulum (ER)

This is found as **rough ER** (with ribosomes) and **smooth ER** (without ribosomes). It is a series of folded internal membranes. Substances are transported in the spaces between the ER. The smooth ER aids the synthesis and transport of lipids.

rough endoplasmic reticulum

smooth endoplasmic reticulum

Golgi body

Look for vesicles 'pinching off' the main Golgi sacs.

This is a series of flattened sacs, each separated from the cytoplasm by a membrane. The **Golgi body** is a packaging system where important chemicals become membrane wrapped, forming **vesicles**. The vesicles become detached from the main Golgi sacs, enabling the isolation of chemicals from each other in the cytoplasm. The Golgi body aids the production and secretion of many proteins, carbohydrates and glycoproteins. Vesicle membranes merge with the plasma membrane to enable secretions to take place.

Golgi body

Lysosomes

These are **specialised vesicles** because they contain **digestive enzymes**. The enzymes have the ability to break down proteins and lipids. If the enzymes were free to react in the cytoplasm then cell destruction would result.

lysosome

microvilli

Microvilli

Microvilli are small projections from the cell membrane of certain cells. They increase the surface area of the cell many times. This allows substances to be absorbed much faster across the cell membrane. Examples of cells that have microvilli are the cells lining the villi in the small intestine.

The plant cell and its organelles

All of the structures described for animal cells are also found in plant cells (except the centrioles). Additionally there are three extra structures shown in the diagram below.

> Did you spot the three extra structures in the plant cell? Remember that a root cell under the soil will not possess chloroplasts. Nor does every plant cell above the soil have chloroplasts.

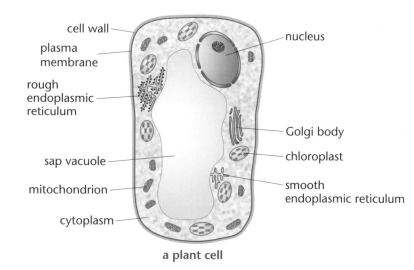

a plant cell

Cell wall

> Cells other than plant cells can have cell walls, e.g. bacteria have polysaccharides other than cellulose.

Around the plasma membrane of plant cells is the cell wall. This is secreted by the cell and consists of cellulose microfibrils embedded in a layer of calcium pectate and hemicelluloses. Between the walls of neighbouring cells calcium pectate cements one cell to the next in multi-cellular plants. Plant cell walls provide a rigid support for the cell but allow many substances to be imported or exported by the cell. The wall allows the cell to build up an effective hydrostatic skeleton. Some plant cells have a cytoplasmic link which crosses the wall. These links of cytoplasm are known as **plasmodesmata**.

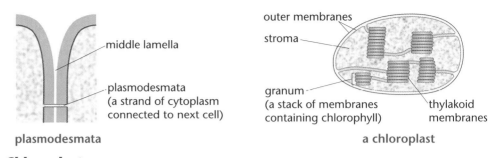

plasmodesmata

a chloroplast

middle lamella

plasmodesmata
(a strand of cytoplasm
connected to next cell)

outer membranes

stroma

granum
(a stack of membranes
containing chlorophyll)

thylakoid
membranes

Chloroplasts

These enable the plant to photosynthesise, making glucose. Each consists of an outer covering of two membranes. Inside are more membranes stacked in piles called **grana**. The membranes enclose a substance vital to photosynthesis, **chlorophyll**. Inside the chloroplast is a matrix known as the **stroma** which is also involved in photosynthesis.

Sap vacuole

This is a large space in a plant cell, containing chemicals such as glucose and mineral ions in water. This solution is the sap. It is surrounded by a membrane known as the **tonoplast**. It is important that a plant cell contains enough water to maintain internal hydrostatic pressure. When this is achieved the cell is turgid, having maximum hydrostatic strength.

Prokaryotic and eukaryotic cells

AQA 1.3

Organisms can be classified into two groups, **prokaryotic** or **eukaryotic** according to their cellular structure.

> Prokaryotic cells are characteristic of two groups of organisms, bacteria and blue-green algae. Prokaryotic cells are less complex than eukaryotic ones and are considered to have evolved earlier.
>
> **KEY POINT**

The table below states similarities and differences between the two types of organism.

In an examination you will often be given a diagram of a cell from an organism you have not seen before. This is not a problem! The examiners are testing your recognition of the organelles found in typical prokaryotic and eukaryotic organisms.

		Prokaryotic cells	*Eukaryotic cells*
Kingdom		Prokaryotae	Protoctista, Fungi, Animalia, Plantae
Organelles	1	small ribosomes	large ribosomes
	2	DNA present but there is no nuclear membrane	DNA is enclosed in a membrane i.e. has nucleus, mitochondria, (Golgi body vesicles and ER are present)
	3	cell wall present consisting of mucopeptides	cell walls present in plant cells – cellulose cell walls present in fungi – chitin
	4	if cells have flagellae there is no 9+2 microtubule arrangement	if cells have flagellae there is a 9+2 microtubule arrangement

Progress check

1 Describe the function of each of the following cell organelles:

nucleus   Golgi body   mitochondria
ribosomes   cell (plasma) membrane

2 Give **three** structural differences between a plant and animal cell.

2 A plant cell has a cellulose cell wall, chloroplasts, and a sap vacuole lined by a tonoplast.
Cell (plasma) **membrane** – gives physical protection to the outside of a cell, allows the import and export of *selected chemicals*.
Golgi **body** – is a packaging system where chemicals become membrane wrapped, forming vesicles
ribosomes – aid the manufacture of proteins, being the site where mRNA meets tRNA so that amino acids are bonded together
mitochondria – release energy during aerobic respiration
1 **nucleus** – mRNA is produced in the nucleus with the help of DNA

2.2 Isolation of cell organelles

Cell fractionation

AQA ▶ 1.3

Occasionally it is necessary to isolate organelles to investigate their structure or function, e.g. mitochondria could be used to investigate aerobic respiration away from the cell's internal environment. **Cell fractionation** consists of two processes: **homogenisation** followed by **differential centrifugation**. Cell fractionation depends on the different densities of the organelles.

Technique

Cells are kept:

- cool at around 5°C *(this slows down the inevitable autolysis, destruction by the cell's own enzymes)*
- in an isotonic solution, i.e. equal concentration of substances inside and outside of the cell membrane *(this ensures that the organelles are not damaged by osmosis and can still function)*
- at a specific pH by a buffer solution *(this ensures that the organelles can still function, as they are kept in suitable conditions)*.

Homogenisation

The cells are homogenised in either a pestle homogeniser or a blender. This breaks the cells up releasing the organelles and cytoplasm. Many organelles are not damaged at all by this process. At this stage there is a suspension of mixed organelles.

pestle homogeniser   blender

Differential centrifugation

Equal amounts of homogenised tissue are poured into the tubes of an **ultracentrifuge**. This instrument spins the cell contents at a force many times greater than gravity. Organelle separation by this technique is **density dependent**.

First spin

600 g means 600 times the force of gravity. If a plant cell was centrifuged at around this speed then chloroplasts would be contained in the sediment as well as nuclei.

The sample is spun at 600 g for 10 minutes. **Nuclei**, the organelles of greatest density, collect in the **sediment** at the base of the tube. All other cell contents are

The principle of pouring off the supernatant to leave the pure sediment behind can be repeated at the end of each spin. In this way the main organelles can be isolated.

Note that it is much easier to obtain nuclei, because they are isolated in the first spin. Ribosomes are isolated at the final spin.

found in the **supernatant** (the fluid above the sediment). Pouring off the supernatant leaves the sediment of relatively pure nuclei.

Second spin

The remaining supernatant fluid is spun at 10 000 to 20 000 g for a further 20 minutes. **Mitochondria**, the most dense of the remaining organelles, collect in the sediment. All other cell contents are found in the **supernatant**.

Third spin

The remaining supernatant fluid is spun at 100 000 g for a further 60 minutes. In this sediment are fragments of endoplasmic reticulum and ribosomes. Other cell contents, e.g. cytoplasm and proteins, remain in the supernatant.

Using this technique the organelles are isolated. By providing them with suitable conditions they remain active for a time and can be used in investigations, e.g. chloroplasts given isotonic solution, suitable (warm) temperature, light, carbon dioxide and water will continue to photosynthesise.

Light and electron microscopy

AQA ▶ 1.3

Microscopes **magnify** the image of a specimen to enable the human eye to see minute objects not visible to the naked eye. **Resolution** of a microscope is the ability to distinguish between two objects as separate entities. At low resolution only one object may be detected. At high resolution two distinct objects are visible. At high resolution the image of such a specimen would show considerable detail.

The light microscope has limited resolution ($0.2\,\mu m$) due to the wavelength of light so that organelles such as mitochondria, although visible, do not have clarity. Electron microscopes have exceptional resolution. The transmission electron microscope has a high resolution (0.2 to $0.3\,nm$). This enables even tiny organelles to be seen.

The light microscope

This type of microscope uses white light to illuminate a specimen. The light is focused onto the specimen by a **condensing lens**. The specimen is placed on a microscope slide which is clipped onto a platform, known as the **stage**. The image is viewed via an eyepiece or ocular lens. Overall magnification of the specimen depends on the individual magnification of the eyepiece lens and objective lens. For example, if a specimen is being observed with an eyepiece × 10 and an objective lens of × 40, then the image is 400 times the true size of the specimen.

> Many specimens need **staining** with chemicals so that tissues and, perhaps, organelles can be seen clearly, e.g. methylene blue is used to stain nuclei. Sometimes more than one stain is used, e.g. in differential staining, so that sub-cellular parts contrast with each other.
>
> **KEY POINT**

To decide which microscope is suitable use the table below.

light microscope

transmission electron microscope

The electron microscope

This uses an electron stream which is directed at the specimen. The **transmission electron microscope (TEM)** has extremely high magnification and resolution properties. Specimens are placed in a vacuum within the microscope, to ensure the electrons do not collide with air molecules and distort the image. **Stains** such as **osmium** and **uranium** salts are used to make organelles distinct. These salts are absorbed by organelles and membranes differentially, e.g. the nuclear membrane absorbs more of the salts than other parts. In this way the nuclear membrane becomes more dense. When the electron beam hits the specimen, electrons are unable to pass through this dense membrane. The membrane shows up as a dark shadow area on the image, because it is in an electron shadow. Cytoplasm allows more electrons to pass through. When these electrons hit the fluorescent screen visible light is emitted. In the scanning electron microscope electrons bounce off the surface of specimens. This provides a 3D image.

Artefacts

When microscopic specimens are prepared there are often several chemical and physical procedures. Often the material is dead so changes from the living specimen are expected. Microscopic material should be analysed with care because there may have been some artificial change in the material during preparation, e.g. next to some cells a student might see a series of small circles. They look like eggs but are merely air bubbles. These are **artefacts**; structures alien to the material which should not be interpreted as part of the specimen.

Microscopic measurement

It is sometimes necessary to measure microscopic structures. There are two instruments needed for this process, a **graticule** and **stage micrometer**.

graticule

stage micrometer

Calibration and measurement technique

- Put a graticule into the eyepiece of a microscope.
- Look through the eyepiece lens and the graticule line can be seen.
- Put a stage micrometer on the microscope stage.
- Look through the eyepiece lens.
- Line up the ruled line of the graticule with the ruled line of the stage micrometer.
- Calibrate the eyepiece by finding out the number of eyepiece units (e.u.) equal to one stage unit (s.u.); each is 0.01 mm.
- If three eyepiece units equal 1 stage unit, then 1 eyepiece unit is equal to 0.01/3 (0.0033 mm).
- Take away the stage micrometer and replace it with a specimen.
- Measure the dimensions of the specimen in terms of eyepiece units.

Important! Every time you change the objective lens, e.g. move from low power to high power, recalibration is necessary. In an examination this may be tested. Many candidates forget to recalibrate. Don't miss the mark!

Example

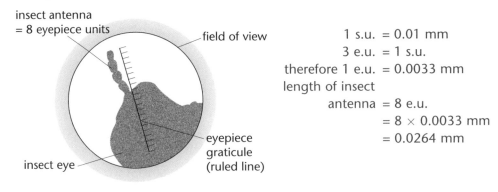

insect antenna
= 8 eyepiece units

field of view

insect eye

eyepiece graticule (ruled line)

1 s.u. = 0.01 mm
3 e.u. = 1 s.u.
therefore 1 e.u. = 0.0033 mm
length of insect
    antenna = 8 e.u.
         = 8 × 0.0033 mm
         = 0.0264 mm

Progress check

1 Describe how the width of an insect's leg could be measured using a microscope, graticule and stage micrometer.

2 A student measured the insect leg but considered that a higher magnification was needed to improve accuracy. What is the significance of changing magnification to the measuring technique?

1 Put a graticule into the eyepiece of a microscope; put a stage micrometer on the microscope stage; look through the eyepiece lens; line up the ruled line of the graticule with the ruled line of the stage micrometer; calibrate the eyepiece by finding out the number of eyepiece units equal to one stage unit; a stage unit is a known length, so an eyepiece unit length can be calculated; take away the stage micrometer and replace with specimen; measure width of insect leg in eyepiece units.

2 Recalibration is needed for each new magnification.

2.3 Specialisation of cells

After studying this section you should be able to:

- *understand cell specialisation and how cells aggregate into tissues and organs*
- *recall a range of cell adaptations*

LEARNING SUMMARY

The earlier parts of this chapter described the structure and function of generalised animal and plant cells. Their features are found in many unicellular organisms where all the life-giving processes are carried out in one cell. Additionally many multicellular organisms exist. A few show no specialisation and consist of repeated identical cells, e.g. Volvox, a colonial alga. Most multicellular organisms exhibit **specialisation**, where different cells are adapted for specific roles.

Cell adaptations

AQA 2.6

Some important features:

Red blood cell
- no nucleus
- high surface area
- contains haemoglobin which has an affinity for oxygen

White blood cell (neutrophil)
- can surround foreign cells and debris
- produces enzymes to digest the foreign material

Epithelial cell (e.g. squamous cell)
- very thin
- allows exchange of chemicals

Plant palisade cell
- chloroplasts for photosynthesis
- chloroplasts can move to absorb more light
- contains chlorophyll which absorbs light
- sap vacuole stores important chemicals

Sperm cell
- flagellum for swimming
- numerous mitochondria for ATP production
- an acrosome containing enzymes to digest way into female gamete

Root hair cell
- projection of cell to increase surface area for water absorption and anchorage

Guard cell
- unevenly thickened cell walls so cells bend when turgid
- chloroplasts to provide energy for uptake of minerals and hence water.

red blood cell

neutrophil

epithelial cell

palisade, mesophyll cell of leaf

sperm cell

root hair cell

guard cells

Tissues, organs and systems

AQA 2.6

A **tissue** is a collection of similar cells, derived from the same source, all working together for a specific function, e.g. palisade cells of the leaf which photosynthesise or the smooth muscle cells of the intestine which carry out peristalsis.

An **organ** is a collection of tissues which combine their properties for a specific function, e.g. the stomach includes: smooth muscle, epithelial lining cells, connective tissue, etc. Together they enable the stomach to digest food.

A range of tissues and organs combine to form a **system**, e.g. the respiratory system.

In multicellular organisms specific groups of cells are specialised for a particular role. This increased efficiency helps the organism to have better survival qualities in the environment.

The photomicrograph below shows some of the cells which are part of a bone.

Haversian canal containing blood vessels and nerves

Bone cells which secrete the minerals which harden the bone

Combinations of cells each contribute their specific adaptations to the overall function of an organ. Compact bone, spongy bone and articular cartilage all have distinct but vital qualities.

Sample question and model answer

The electron micrograph below shows a lymphocyte which secretes antibodies. Antibodies are proteins.

Analyse electron micrographs carefully. Learn the typical structures of all the organelles and then you will be prepared.

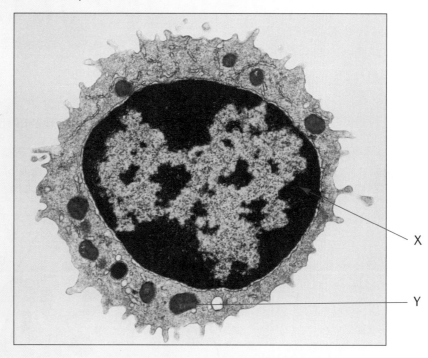

(a)

(i) Name the organelles X and Y.   [2]

X = nucleus
Y = mitochondrion

(ii) The cell was stained with uranium salts in preparation for a transmission electron microscope. Explain how this stain caused the nucleus to show a dark shade compared to the light shade of the cytoplasm.   [4]

There are many different stains. You need to know the principle of how they work, but not specific names.

The stain was taken up by the nucleus more than the cytoplasm; the electrons could not pass through the stained (dense) parts of the nucleus so the dark nucleus parts on the screen are in electron shade. Electrons pass through the cytoplasm and cause light emission (fluorescence) at the screen.

(b)

(i) Given a piece of liver how would you isolate mitochondria from the cells?   [7]

Put the liver in isotonic solution;
homogenise the liver or grind with a pestle and mortar;
filter the homogenate through muslin layers to remove cell debris;
put in a centrifuge; spin at 500 – 600 g for 10 minutes;
discard the pellet or sediment; centrifuge the supernatant;
spin at 10 000 – 20 000 g for 20 minutes;
mitochondria are now in the pellet or sediment.

Note that there are 7 marks maximum for this question. Any 7 of the responses shown would achieve maximum credit. The ideal answer given shows 9 potential creditable points.

(ii) Why is it important to keep fresh liver cells at a temperature of around 5°C during the preparation of the sample?   [2]

They keep alive for a longer period since a low temperature slows down the action of the enzymes which break down the mitochondria; i.e. prevents or slows down cell autolysis (self-digestion of the cells).

Practice examination questions

1 The diagram shows the structure of a cell surface membrane.

(a) Name molecules A and B. [2]

(b) Describe one role of molecule B. [1]

(c) Explain why the structure of molecule A means that the membrane forms a bilayer [2]

2 The diagram below shows an electron micrograph of a cell.

(a) Name the parts labelled **A**, **B**, **C**, **D** and **E**. [5]

(b) The magnification of this diagram is 10 000.
Work out the actual diameter of the nucleus.
Give your answer in micrometers. [3]

(c) The cell is a liver cell. It contains many mitochondria.

Explain why there are so many mitochondria in each liver cell. [2]

3 The diagram below shows a bacterium.

(a) Describe **two** features, visible in the diagram, which show that the bacterium is a prokaryotic organism. [2]

(b) Name **two** organelles from a human cell which show that it is a eukaryotic organism. [2]

Practice examination questions *(continued)*

4 The diagram below shows the structure of a transmission electron microscope (TEM).

transmission electron microscope

(a) Name lens X and lens Y. [2]

(b) Why is it necessary for the specimen to be put in a vacuum? [1]

(c) Occasionally an image seen when using the electron microscope shows an item not present in the living organism.

    (i) What name is given to this type of item? [1]

    (ii) How should the presence of the item be interpreted? [1]

Chapter 3
Enzymes

The following topics are covered in this chapter:

- *Enzymes in action*
- *Inhibition of enzymes*

- *Enzymes and digestion*
- *Applications of enzymes*

3.1 Enzymes in action

After studying this section you should be able to:

- understand the role of the active site and the enzyme–substrate complex in enzyme action
- understand how enzymes catalyse biochemical reactions by lowering activation energy
- understand the factors which affect the rate of enzyme catalysed reactions

LEARNING SUMMARY

How enzymes work

AQA   1.1

Living cells carry out many biochemical reactions. These reactions take place **rapidly** due to the presence of enzymes. All enzymes consist of **globular proteins** which have the ability to 'drive' biochemical reactions. Some enzymes require additional non-protein groups to enable them to work efficiently, e.g. the enzyme dehydrogenase needs a coenzyme NAD to function. Most enzymes are contained within cells but some may be released and act extracellularly.

> The tertiary folding of polypeptides are responsible for the special shape of the active site.

> **KEY POINT**
> The ability of an enzyme to function depends on the specific shape of the protein molecule. The intricate shape created by polypeptide folding (see page 22) is a key factor in both theories of enzyme action.

Lock and key theory

> In an examination the lock and key theory is the most important model to consider. Remember that both catabolic and anabolic reactions may be given.

- Some part of the enzyme has a cavity with a precise shape (**active site**).
- A substrate can fit into the active site.
- The active site (lock) is exactly the correct shape to fit the substrate (key).
- The substrate binds to the enzyme forming an **enzyme–substrate complex**.
- The reaction takes place rapidly.
- Certain enzymes break a substrate down into two or more products (**catabolic** reaction).
- Other enzymes bind two or more substrates together to assemble one product (**anabolic** reaction).

> metabolic reaction
> = anabolic + catabolic
>   reaction   reaction
> Remember that metabolism is a summary of **build up** and **break down reactions**.

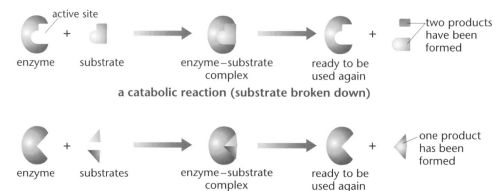

enzyme   substrate   enzyme–substrate complex   ready to be used again   two products have been formed

a catabolic reaction (substrate broken down)

enzyme   substrates   enzyme–substrate complex   ready to be used again   one product has been formed

an anabolic reaction (substrates used to build a new molecule)

Induced fit theory

- The active site is a cavity of a particular shape.
- Initially the active site is not the correct shape in which to fit the substrate.
- As the substrate approaches the active site, the site changes and this results in it being a perfect fit.
- After the reaction has taken place, and the products have gone, the active site returns to its normal shape.

enzyme   substrate

The shape of the active site changes as the substrate approaches.

enzyme–substrate complex

Active site is a perfect shape for the substrate.

products

Lowering of activation energy

Every reaction requires the input of energy. Enzymes reduce the level of activation energy needed as shown by the graph.

The higher the activation energy the slower the reaction. An enzyme reduces the amount of energy required for a biochemical reaction. When an enzyme binds with a substrate the available energy has a greater effect and the rate of catalysis increases. The conditions which exist during a reaction are very important when considering the rate of progress. Each of the following has an effect on the rate:

- concentration of substrate molecules
- concentration of enzyme molecules
- temperature
- pH.

You may be questioned on the factors which affect the rate of reaction. Less able candidates tend to remember just one or two factors. Learn all four factors here and achieve a higher grade!

Look out for questions which show the rate of reaction graphically. The examiners often test your understanding of limiting factors (see practice question on page 47).

What is the effect of enzyme and substrate concentration?

When considering the rate of an enzyme catalysed reaction the proportion of enzyme to substrate molecules should be considered. Every substrate molecule fits into an active site, then the reaction takes place. If there are more substrate molecules than enzyme molecules then the number of active sites available is a **limiting factor**. The **optimum rate** of reaction is achieved when all the active sites are in use. At this stage if more substrate is added, there is no increase in rate of product formation. When there are fewer substrate molecules than enzyme molecules the reaction will take place very quickly, as long as the conditions are appropriate.

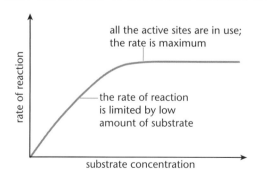

Remember that particles in liquids (and gases) are in constant random motion, even though we cannot see them.

How does temperature affect the rate of an enzyme catalysed reaction?

- **Heat energy** reaching the enzyme and substrate molecules causes them to **increase random movement**.
- The greater the heat energy the more the molecules move and so the more often they **collide**.
- The more **collisions** there are the greater the chance that substrates will fit into an **active site**, up to a specific temperature.
- At the **optimum** temperature of an enzyme, the reaction rate is maximum.
- Heat energy also affects the shape of the active site, the active site being the correct shape at the optimum temperature.
- At temperatures above optimum, the rate of reaction decreases because the active site begins to distort.
- Very high temperature causes the enzyme to become **denatured**, i.e. bonding becomes irreversibly changed and the active site is **permanently damaged**.
- At very high temperatures, the number of collisions is correspondingly high, but without active sites no products can be formed.
- At lower temperatures than the optimum, the rate of the reaction decreases because of reduced enzyme–substrate collisions.

Most enzymes have an optimum temperature of between 30°C and 40°C, but there are many exceptions. An example of this is shown by some bacteria that live at high temperatures in hot volcanic springs.

It is interesting to consider that some micro organisms can spoil ice-cream in a freezer whereas a different micro organism, with different enzymes, can decompose grass in a 'steaming' compost heap.

How does pH affect the rate of an enzyme catalysed reaction?

The **pH** of the medium can have a direct effect on the bonding responsible for the **secondary and tertiary structure** of enzymes. If the active site is changed then enzyme action will be affected. Each enzyme has an optimum pH.

- Many enzymes work best at **neutral** or **slightly alkaline** conditions, e.g. salivary amylase.
- Pepsin works best in **acid** conditions around pH 3.0, as expected considering that the stomach contains hydrochloric acid.

Remember that other factors affect an enzyme catalysed reaction:
- substrate concentration
- enzyme concentration
- temperature.

Each can be a limiting factor.

For the two enzyme examples above, the active sites are ideally shaped at the pHs mentioned. An inappropriate pH, often acidic, can change the active site drastically, so that the substrate cannot bind. The reaction will not take place. On most occasions the change of shape is not permanent and can be returned to optimum by the addition of an alkali.

Progress check

How does temperature affect the rate of the reaction by which protein is changed to polypeptides by the enzyme pepsin, in the human stomach?

$$\text{protein} \xrightarrow{\text{pepsin}} \text{polypeptides}$$

- Heat energy causes the enzyme and substrate molecules to increase random movement, increasing the chance of collision.
- At 37°C (optimum temperature) there is a greater chance that the protein will fit into an active site, so the production of polypeptides is at maximum rate.
- At 37°C (optimum temperature) the shape of the active site is best suited to fit the protein.
- At temperatures higher than 37°C the rate of reaction decreases because the active site begins to distort.
- Very high temperature causes the pepsin to become denatured, i.e. bonding has been irreversibly changed and the active site is permanently damaged.
- At very high temperatures the number of collisions is correspondingly high, but without active sites no polypeptides can be formed.
- At lower temperatures than optimum the rate of reaction decreases because of reduced enzyme–substrate collisions.

3.2 Inhibition of enzymes

After studying this section you should be able to:

- *understand the action and effects of competitive and non-competitive inhibitors*

LEARNING SUMMARY

What are inhibitors?

AQA ▶ 3.1.1

Certain chemicals can slow down or stop enzyme catalysed reactions. These chemicals are called **inhibitors**.

Sometimes these chemicals are substances that are naturally occurring inside cells. They may be used to regulate the rate of enzyme controlled reactions. Other inhibitors may be poisons or medicines.

Competitive inhibitors

- These are molecules of **similar shape** to the normal substrate and are able to bind to the active site.
- They do not react within the active site, but leave after a time without any product forming.
- The enzymic reaction is **reduced** because while the inhibitor is in the active site, **no substrate can enter**.
- Substrate molecules **compete** for the active site so the rate of reaction decreases.
- The higher the proportion of competitive inhibitor the slower the rate of reaction.

substrate

competitive inhibitor

substrate cannot enter active site

the inhibitor binds with the active site

substrate may now enter

Non-competitive inhibitors

Some enzymes have two sites, the active site and one other. An **allosteric** molecule fits into the alternative site. Here it changes the shape of the active site. This can stimulate the reaction if the active site becomes a better shape (**allosteric activation**). It can also inhibit if the active site becomes an inappropriate shape (**allosteric inhibition**).

- These are molecules which bind to some part of an enzyme other than the active site.
- They have a different shape to the normal substrate.
- They change the shape of the active site which no longer allows binding of the substrate.
- Some substrate molecules may reach the active site before the non-competitive inhibitor.
- The rate of reaction is reduced.
- Finally they leave their binding sites, but substrate molecules do not compete for these, so they have a greater inhibitory effect.

non-competitive inhibitor

substrate

binding site

active site has changed

substrate has opportunity to enter

The graph below shows the relative effects of competitive and non-competitive inhibitors, compared to a normal enzyme catalysed reaction.

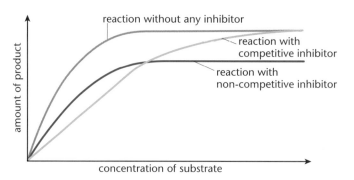

Sometimes an inhibitor will not leave the enzyme once it has bound with it. It stays permanently attached. This is called an **irreversible inhibitor**. Inhibitors that do leave are called **reversible**.

Progress check

1 Explain why a non-competitive inhibitor does not need to have the same shape as the substrate. Give a reason for your answer.

2 How does a non-competitive inhibitor reduce the rate of an enzyme catalysed reaction?

1 No. Non-competitive inhibitors have a different shape to the normal substrate. They bind to some part of an enzyme other than the active site.
2 They change the shape of the enzyme's active site which is less suitable for the binding of the substrate, so the rate of reaction decreases.

3.3 Enzymes and digestion

After studying this section you should be able to:

- *recall the structure of the human digestive system*
- *recall the sites of secretion and action of enzymes involved in carbohydrate digestion*

L E A R N I N G   S U M M A R Y

The human digestive system

AQA   1.2

Humans feed **heterotrophically** by taking in complex organic substances. This is called **ingestion**. These substances are then broken down into smaller, soluble molecules by **digestion**.

Once digestion has occurred, the molecules can be **absorbed** into the blood stream. They can then be built up or **assimilated** into the various complex molecules that the body requires.

The human digestive system is made up of a long tube that starts at the mouth and ends at the anus. Along its length, different secretions are added to the food either from the lining of the system or from associated glands.

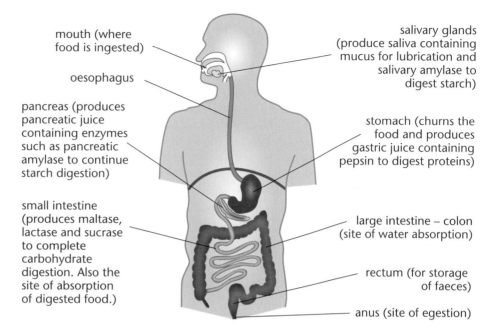

mouth (where food is ingested)

oesophagus

pancreas (produces pancreatic juice containing enzymes such as pancreatic amylase to continue starch digestion)

small intestine (produces maltase, lactase and sucrase to complete carbohydrate digestion. Also the site of absorption of digested food.)

salivary glands (produce saliva containing mucus for lubrication and salivary amylase to digest starch)

stomach (churns the food and produces gastric juice containing pepsin to digest proteins)

large intestine – colon (site of water absorption)

rectum (for storage of faeces)

anus (site of egestion)

Although different regions of the digestive system are specialised for different jobs, they are all built on a similar pattern. The diagram shows the different layers in the wall of the digestive system.

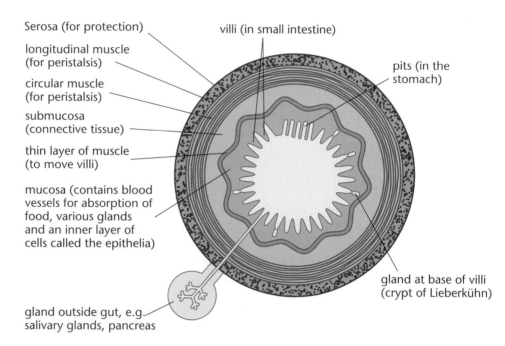

Serosa (for protection)

longitudinal muscle (for peristalsis)

circular muscle (for peristalsis)

submucosa (connective tissue)

thin layer of muscle (to move villi)

mucosa (contains blood vessels for absorption of food, various glands and an inner layer of cells called the epithelia)

villi (in small intestine)

pits (in the stomach)

gland at base of villi (crypt of Lieberkühn)

gland outside gut, e.g. salivary glands, pancreas

Digestion of carbohydrates

AQA   1.2

The human digestive enzymes are **hydrolases**. Each time a molecule is broken down water is required during the reaction. The table below includes some human digestive enzymes that are involved in carbohydrate digestion.

Site in alimentary canal	Secretion	Enzyme	Substrate	Product
mouth	saliva	amylase	starch	maltose
duodenum	pancreatic juice	amylase	starch	maltose
small intestine	intestinal juice	maltase	maltose	glucose
		sucrase	sucrose	glucose & fructose
		lactose	lactose	glucose & galactose

3.4 Applications of enzymes

LEARNING SUMMARY

After studying this section you should be able to:

- *describe and understand the use of lactase in producing lactose-free products*

Lactose removal from milk

AQA ▶ 1.2

We currently live in a golden age of biotechnology and enzymes are now used in a variety of applications in the home, in medicine and in industry.

Lactose removal from milk is necessary because some people have allergic reactions to lactose, the sugar in milk. **Lactase** is added to change the lactose to glucose and galactose. As a result, lactose-free milk is available to consumers.

Sample question and model answer

The graph below shows the rate of an enzyme catalysed reaction, with and without a **non-competitive inhibitor**, at different substrate concentrations.

(a) Give **one** piece of evidence from the graph which shows that the inhibitor was non-competitive rather than competitive. [1]

> When a non-competitive inhibitor is present the maximum rate of the inhibitor-free reaction is not reached.

A competitive inhibitor would still allow the maximum rate at high substrate concentration.

(b) Explain why the rates of the inhibited and non-inhibited reactions were very similar up to a substrate concentration of 20 μmol cm⁻³. [2]

> Before a concentration of 20 μmol cm⁻³
> - in both reactions substrate molecules are similarly successful in reaching active sites
> - few inhibitor molecules have become bound to the alternative sites on the enzyme
> - few active sites have been changed so are still available for the substrate.

Remember that the non-competitive inhibitor binds to a different part of an enzyme and NOT the active site. It still causes a change in the active site which cannot then bind with the substrate.

(c) Give **two** factors that are needed to be kept constant when investigating both the inhibited and non-inhibited reactions. [2]

> - temperature
> - pH

Practice examination questions

1   Describe each of the following pairs to show that you understand the main differences between them:

(a) the lock and key enzyme theory **and** the induced fit enzyme theory

(b) reversible **and** irreversible inhibitors.   [4]

2   In terms of the tertiary structure of the enzyme, explain why amylase can break down starch but has no effect on a lipid.   [3]

3   Bacterial α-amylase works best at around 80°C.

(a) Name the substrate which it breaks down.   [1]

(b) Why is this enzyme described as a thermostable enzyme?   [1]

4   The diagram represents an enzyme and its substrate.

(a) Referring to information in the diagram explain the activity of this enzyme in terms of the **induced fit theory**.   [2]

(b) Molecule X is a non-competitive inhibitor. Explain how this inhibitor has an effect on the structure and function of the enzyme.   [3]

5   The diagram below shows a long polypeptide.

(a) The carboxylic acid is found at one end of the polypeptide. Which group is found at the other end?   [1]

(b) Pepsin breaks down polypeptides by breaking bonds in the centre of the molecule. This speeds up the subsequent digestion of the polypeptide. Explain how.   [2]

Chapter 4

Exchange

The following topics are covered in this chapter:

- *The cell surface membrane*
- *The movement of molecules in and out of cells*
- *Gaseous exchange in humans*
- *Gaseous exchange in other organisms*

4.1 The cell surface membrane

After studying this section you should be able to:

- understand the importance of surface area to volume ratio
- recall the fluid mosaic model of the cell surface (plasma) membrane

LEARNING SUMMARY

How important is the surface area of exchange surfaces?

AQA A ▶ 2.7

Unicellular organisms like amoeba have a very high **surface area to volume ratio**. All chemicals that are needed can pass into the cells directly and all waste can pass out efficiently. Organisms which have a high surface area to volume ratio have no need for special structures like lungs or gills.

Nutrients and oxygen passing into an organism are rapidly used up. This limits the ultimate size to which a unicellular organism can grow. If vital chemicals did not reach all parts of a cell then death would be a consequence.

A unicellular organism may satisfy all its needs by direct diffusion. However, in larger organisms cells join to adjacent ones and surfaces exposed for exchange of substances are reduced. The larger an organism the lower its surface area to volume ratio. For this reason many multicellular organisms have specially adapted exchange structures.

Fluid mosaic model of the cell surface (plasma) membrane

AQA ▶ 1.3

Ultimately the exchange of substances takes place across the cell surface membrane. This must be selective, allowing some substances in and excluding others. The cell membrane consists of a bilayer of phospholipid molecules (see page 20–21). Each phospholipid is arranged so that the hydrophilic (attracts water) head is facing towards either the cytoplasm or the outside of the cell. The hydrophobic (repels water) tails meet in the middle of the membrane. Across this expanse of phospholipids are a number of protein molecules. Some of the proteins (intrinsic) span the complete width of the membrane, some proteins (extrinsic) are partially embedded in the membrane.

Remember that plant cells have a cellulose cell wall. This gives physical support to the cell but is permeable to many molecules. Water and ions can readily pass through.

The fluid mosaic model of the cell membrane

upper surface of cell membrane protein

phospholipid head

outside of cell

protein channel to transport specific substances into cell

phospholipid bilayer

inside of cell

protein molecule

hydrophobic tail
hydrophilic head

phospholipid molecule

Functions of cell membrane molecules

The term 'fluid mosaic' was given to the cell membrane because of the dynamic nature of the component molecules of the membrane. Many of the proteins seem to 'float' through an apparent 'sea' of phospholipids. Few molecules are static. The fluidity of the membrane is controlled by the quantity of cholesterol molecules. These are found inbetween the tails of the phospholipids.

Phospholipid

Small lipid-soluble molecules pass through the membrane easily because they dissolve as they pass through the phospholipid's bilayer. Small uncharged molecules also pass through the bilayer.

small lipid-soluble molecules pass through

Channel proteins (ion gates)

Larger molecules and charged molecules can pass through the membrane due to channel proteins. Some are adjacent to a receptor protein, e.g. at a synapse a transmitter substance binds to a receptor protein. This opens the channel protein or ion gate and sodium ions flow in.

Not all channel proteins need a receptor protein.

When the molecule binds to a receptor molecule it is similar to a substrate binding with an enzyme's active site. On this occasion the receptor site is the correct shape.

transmitter substance

receptor protein

Na^+ ion gate open

Na^+

Carrier protein molecule

Some molecules which approach a cell may bind with a carrier protein. This has a site which the incoming molecule can bind to. This causes a change of shape in the carrier protein which deposits the molecule into the cell cytoplasm.

once in position the molecule changes the shape of the carrier protein

the site gives up the molecule on the inside of the cell

carrier protein

Recognition proteins

These are extrinsic proteins, some having carbohydrate components, which help in cell recognition (cell signalling) and cell interaction, e.g. foreign protein on a bacterium would be recognised by white blood cells and the cell would be attacked. The combination of a protein with a carbohydrate is called a **glycoprotein**. The carbohydrate chains are only on the outside of the cell membrane and are called the **glycocalyx**.

White blood cells continually check the proteins on cell membranes. Those recognised as 'self' are not attacked, whereas those which are not 'self' are attacked.

carbohydrate

recognition protein

> **KEY POINT**
>
> The cell surface membrane is the key structure which forms a barrier between the cell and its environment. Nutrients, water and ions must enter and waste molecules must leave. Equally important is the exclusion of dangerous chemicals and inclusion of vital cell contents. It is no surprise that the cell makes further compartments within the cell using membranes of similar structure to the cell surface membrane. High temperatures can destroy the structure of the cell membrane and then it loses its ability to contain the cell contents.

4.2 The movement of molecules in and out of cells

After studying this section you should be able to:

- understand the range of methods by which molecules cross cell membranes
- understand the processes of diffusion, facilitated diffusion, osmosis and active transport

How do substances cross the cell surface membrane?

AQA   1.3

Cells need to obtain substances vital in sustaining life. Some cells secrete useful substances but all cells excrete waste substances. There are several mechanisms by which molecules move across the cell surface membrane.

Diffusion

Note that diffusion is the movement of molecules down a concentration gradient.

Molecules in liquids and gases are in constant random motion. When different concentrations are in contact, the molecules move so that they are in equal concentration throughout. An example of this is when sugar is put into a cup of tea. If left, sugar molecules will distribute themselves evenly, even without stirring. Diffusion is the movement of molecules from where they are in high concentration to where they are in low concentration. Once evenly distributed the *net* movement of molecules stops.

Factors which affect the rate of diffusion

- Surface area.
  the greater the surface area the greater the rate of diffusion
- The difference in concentration on either side of the membrane.
  the greater the difference the greater the rate
- The size of molecules.
  smaller molecules may pass through the membrane faster than larger ones
- The presence of pores in the membrane.
  pores can speed up diffusion
- The width of the membrane.
  the thinner the membrane the faster the rate.

Sometimes the membrane is stated as being selectively permeable, partially permeable or semi-permeable. They all mean the same thing.

Facilitated diffusion

This is a special form of diffusion in which protein carrier molecules are involved. It is much faster than regular diffusion because of the carrier molecules. Each carrier will only bind with a specific molecule. Binding changes the shape of the carrier which then deposits the molecule into the cytoplasm. No energy is used in the process.

selectively permeable membrane

A   B

) water molecule
● solute molecule

water molecules move from B to A

Osmosis

This is the movement of water molecules across a selectively permeable membrane:

- from a lower concentrated solution to a higher concentrated solution
- from where water molecules are at a higher concentration to where they are at a lower concentration
- from a hypotonic solution to a hypertonic solution
- from a hyperosmotic solution to a hypo-osmotic solution
- from an area of higher water potential to lower water potential.

The diagram on the left shows a model of osmosis.

*Remember that osmosis is about the movement of **water molecules**. No other substance moves!*

What is the relationship between water potential of the cell and the concentration of an external solution?

The term 'water potential' is used as a measure of water movement from one place to another in a plant. It is measured in terms of pressure and the units are either kPa (kilopascals) or MPa (megapascals). Water potential is indicated by the symbol ψ *(pronounced psi)*. The following equation allows us to work out the water 'status' of a plant cell.

ψ (cell) water potential (of cell)	=	ψs solute potential (of ions inside cell)	+	ψp pressure potential (of cell wall)	K E Y P O I N T

> Remember that water moves from an area of higher water potential to an area of lower water potential. When a cell at -4 MPa is next to a cell at the less negative value the water moves to the more negative value, i.e. -4 MPa > -6 MPa

concentration of solution outside cell increases

total plasmolysis
- vacuole has almost disappeared
- minimum hydrostatic pressure
- also known as flaccid

incipient plasmolysis
- cell membrane begins to leave the cell wall as water is lost

full turgidity
- vacuole maximum volume
- no more water can enter
- maximum hydrostatic pressure
- cell membrane is forced against the cell wall

> Note that pressure potential only has a value **above** zero when the cell membrane **begins** to contact the cell wall. The greater the pressure potential the more the cell wall resists water entry. At turgidity ψs = ψp when net water movement is zero.

Active transport

> Note that active transport is the movement of molecules up a concentration gradient.

In **active transport** molecules move from where they are in lower concentration to where they are in higher concentration. A protein carrier molecule is used. This is **against the concentration gradient** and always **needs energy**. A plant may contain a higher concentration of Mg^{2+} ions than the soil. It obtains a supply by active transport through the cell surface membranes of root hairs. Only Mg^{2+} ions can bind with the specific protein carrier molecules responsible for their entry into the plant. This is also known as active ion uptake, but is a form of active transport. Any process that reduces respiration in cells will reduce active transport, e.g. adding a poison such as cyanide or reducing oxygen availability. This is because energy is needed for the process and this energy is released by respiration.

In the small intestine, glucose is absorbed into the bloodstream indirectly by active transport. Sodium ions are pumped out of the epithelial cells allowing glucose and sodium ions to diffuse into the cell by a co-transporter system.

Endocytosis, exocytosis, pinocytosis and phagocytosis

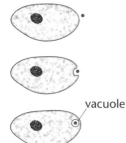

endocytosis

vacuole

Some substances, often due to their large size, enter cells by **endocytosis** as follows:

- the substance contacts the cell surface membrane which indents
- the substance is surrounded by the membrane, forming a vacuole or vesicle
- each vacuole contains the substance and an outer membrane which has detached from the cell surface membrane.

51

When fluids enter the cell in this way this is known as **pinocytosis**. When the substances are large solid particles, this is called **phagocytosis**. Some substances leave the cell in a reverse of endocytosis. Here the membrane of the vacuole or vesicle merges with the cell surface membrane depositing its contents into the outside environment of the cell. This is known as **exocytosis**.

Progress check

1 A plant contains a greater concentration of Fe^{2+} ions than the soil in which it is growing. Name and describe the process by which the plant absorbs the ions against the concentration gradient.

2 Explain the following:
(a) Endocytosis of an antigen by a phagocyte
(b) Exocytosis of amylase molecules from a cell.

1 **Active transport:** molecules move from a lower concentration to a higher concentration; through the cell surface membranes of root hairs; protein carrier molecules in membranes used; energy needed; Fe^{2+} ions can bind with the protein carrier molecules which allow entry into the plant.

2 (a) **Endocytosis:** antigen contacts the cell membrane of the phagocyte; cell membrane surrounds the antigen, forming a vacuole; the vacuole contains the antigen and an outer membrane which has detached from the cell surface membrane.
(b) **Exocytosis:** a vesicle in the cell contains amylase molecules; the vesicle merges with the cell membrane; amylase contents deposited outside of the cell.

4.3 Gaseous exchange in humans

After studying this section you should be able to:

- *understand why organisms need to be adapted for gaseous exchange*
- *explain how the human breathing system brings about ventilation and gaseous exchange*

How are organisms adapted for efficient gaseous exchange?

AQA ▶ 1.4, 2.7

The range of respiratory surfaces in this chapter each have common properties, such as high surface area to volume ratio, one cell thick lining tissue, many capillaries.

The exchange of substances across cell surface membranes has been described. Larger organisms have a major problem in exchange because of their low surface area to volume ratio. Some organisms, like flatworms, have a large surface area due to the shape of their body. Other organisms satisfy their needs by having tissues and organs which have special adaptations for efficient exchange. In simple terms, these structures achieve a very high surface area, e.g. a leaf, and link to the transport system to allow import and export from the organ.

Gaseous exchange in humans

AQA 1.4

You should be aware of the effects of smoking (p90), TB (p93), fibrosis, asthma and emphysema on the function of the lungs.

The diagram below shows the **human gas exchange system**. The alveoli are the site of gaseous exchange and they are connected to the outside air via a system of branching tubes.

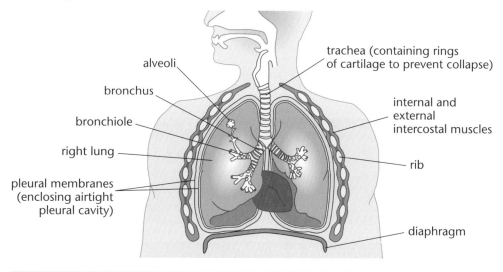

alveoli

bronchus

bronchiole

right lung

pleural membranes (enclosing airtight pleural cavity)

trachea (containing rings of cartilage to prevent collapse)

internal and external intercostal muscles

rib

diaphragm

Ventilation

Drawing air in and out of the lungs involves changes in pressure and volume in the chest. These changes work because the **pleural** membranes form an airtight **pleural cavity**.

Breathing in (inhaling):
1. The external intercostal muscles contract, moving the ribs upwards and outwards.
2. The diaphragm contracts and flattens.
3. Both of these actions will increase the volume in the pleural cavity and so decrease the pressure.
4. Air is therefore drawn into the lungs.

Breathing out (exhaling):
1. The internal intercostal muscles relax and the ribs move down and inwards.
2. The diaphragm relaxes and domes upwards.
3. The volume in the pleural cavity is decreased so the pressure is increased.
4. Air is forced out of the lungs.

Lungs of a mammal

The ventilation mechanism allows inhalation of air, which then diffuses into alveoli to exchange the respiratory gases. Completion of ventilation takes place when gases are expelled into the atmosphere. The diagram on the left shows the structure of the alveoli.

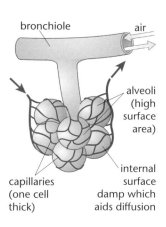

bronchiole

air

alveoli (high surface area)

internal surface damp which aids diffusion

capillaries (one cell thick)

Adaptations of lungs for gaseous exchange

*   Air flows through a trachea (windpipe) supported by cartilage.
*   It reaches the alveoli via tubes known as bronchi and bronchioles.
*   Lungs have many alveoli (air sacs) which have a high surface area.
*   Each alveolus is very thin (diffusion is faster over a short distance).
*   Each alveolus has an inner film of moisture containing a chemical called surfactant. This reduces the surface tension and makes it easier to inflate the lungs.
*   Each alveolus has many capillaries, each one cell thick, to aid diffusion.
*   There are many blood vessels in the lungs to give a high surface area for gaseous exchange and transport of respiratory substances.

The **maximum** volume of air that can be breathed out in one breath is called the vital capacity.

Measuring ventilation

The process of ventilation can be investigated using a device called a **spirometer**. It can measure the volume of air exchanged in a single breath. This is called the **tidal volume**. It can also measure the number of breaths per minute – the **breathing rate**. If these two figures are multiplied together, the result gives the volume of gas exchanged in one minute, the **pulmonary ventilation**.

pulmonary ventilation   =   tidal volume   ×   breathing rate

4.4 Gaseous exchange in other organisms

After studying this section you should be able to:

● *recall where gaseous exchange occurs in a dicotyledonous leaf, the gills of a bony fish and the tracheal system of an insect*
● *show awareness of the adaptations of leaves, gills and tracheoles for efficient gaseous exchange*

LEARNING SUMMARY

A dicotyledonous leaf

AQA ▶ 2.7

The diagram below shows a section through a leaf. Leaves of plants give a high surface area over which exchange takes place. Specialised tissues increase the efficiency of exchange to allow photosynthesis to supply the plant with enough energy-rich carbohydrates.

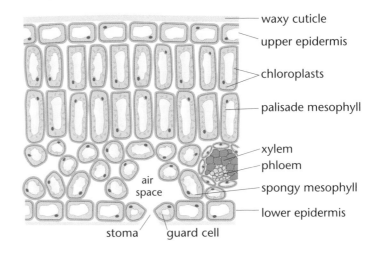

waxy cuticle
upper epidermis
chloroplasts
palisade mesophyll
xylem
phloem
spongy mesophyll
lower epidermis
air space
stoma
guard cell

Adaptations of a leaf for photosynthesis

● A flat, thin blade (lamina) to allow maximum light absorption.
● Cells of the upper epidermis have a waxy cuticle to reflect excess light but allow entry of enough light for photosynthesis.
● Each leaf has many chloroplasts to absorb a maximum amount of light.
● Chloroplasts contain many thylakoid membranes, stacked in grana to give a high surface area to absorb the maximum quantity of light.
● Palisade cells, containing chloroplasts, pack closely together to 'capture' the maximum amount of light.
● Many guard cells open stomata to allow carbon dioxide in and oxygen out during photosynthesis.

- Air spaces in the mesophyll store lots of carbon dioxide for photosynthesis or lots of oxygen for respiration.
- Xylem of the vascular bundles brings water to the leaf for photosynthesis.
- Phloem takes the carbohydrate away from the leaf after photosynthesis.

The gills of a bony fish

AQA ▶ 2.7

gill arch

gill lamella (or primary lamella)

gill plate (or secondary lamella)

direction of blood flow

water current

The ventilation mechanism of a fish allows intake of water, and passes it across the gills. The diagram on the left shows the structures of the gills which allow maximum exchange to take place.

Adaptations of gills for gaseous exchange

- The gills of a bony fish have a very high surface area to volume ratio.
- Gills consist of many flat gill filaments, stacked on top of each other, to give a high surface area for maximum exchange.
- Each gill filament has many gill plates which further increase surface area.
- Gill plates are very thin and full of blood capillaries to aid exchange.
- The gradients of O_2 and CO_2 are kept at a maximum by the **counter-current** flow mechanism. By allowing water to flow over the gills in an opposite direction to blood, maximum diffusion rate is achieved.

All respiratory surfaces are damp to allow effective transport across cell membranes.

Progress check

Describe and explain how the gills of a bony fish are adapted for efficient gaseous exchange.

Each gill consists of many thin gill filaments, stacked on top of each other which give a high surface area to volume ratio, for maximum exchange of gases; each gill filament has many gill plates which further increase surface area; each gill plate is very thin and full of blood capillaries; the gradients of O_2 and CO_2 are kept at a maximum by counter-current flow; water flows over the gills in an opposite direction to blood to achieve maximum diffusion rates.

The tracheal system of insects

AQA ▶ 2.7

An insect is covered by an exoskeleton which is covered in wax and is impermeable. Insects have a series of holes in the cuticle called spiracles. These spiracles lead into a set of tubes called the tracheal system. The larger tubes, called tracheae, branch into smaller tubes called tracheoles. These tiny tubes are permeable and end on the cells of the body. Oxygen can therefore diffuse from the air into the cells. This system helps reduce water loss from the insect.

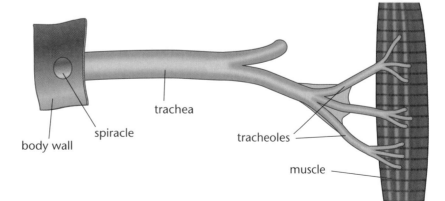

trachea

spiracle

body wall

tracheoles

muscle

Sample question and model answer

The diagram below shows two adjacent plant cells A and B.
The water potential equation is:

$$\psi(\text{cell}) = \psi s + \psi p$$

$\Psi(\text{cell}) = -10\text{MPa}$
$\Psi s = -10\text{MPa}$

$\Psi s = -8\text{MPa}$
$\Psi p = 3\text{MPa}$

cell A   cell B

(a) (i) Calculate the water potential of cell B.   [1]

> Using the equation:
> $\psi(\text{cell}) = \psi s + \psi p$
> $\qquad\quad = -8 \text{ MPa} + 3 \text{ MPa}$
> $\qquad\quad = -5 \text{ MPa}$

In this question you were given the equation. Try to remember it because it could earn you a mark. Additionally, if there were calculations you need the equation to access the other marks.

When given any two of the values you can work out the other.

(ii) Draw an arrow on the diagram to show the direction of water flow. Show how you worked out the direction.   [2]

> The arrow should be drawn from cell B to cell A.
> Direction from -5 MPa to -10 MPa.

(iii) What is the value of the pressure potential (ψp) of cell A?   [1]

> $\psi(\text{cell}) = \psi s + \psi p,$
> $-10 \text{ MPa} = -10 \text{ MPa} + \psi p$
> $\qquad \psi p = -10 \text{ MPa} + 10 \text{ MPa}$
> $\qquad\quad = 0 \text{ MPa}$

Note that from total plasmolysis up to incipient plasmolysis the resistance of the cell, i.e. ψp is zero.

Only when the cell membrane contacts with the wall does it have an effect.

(iv) Name the condition of the cell when $\psi(\text{cell}) = 0$ [1]

> full turgor or fully turgid

(b) Give **one** difference between the following terms: [2]

facilitated diffusion

> molecules move down a gradient

active transport

> energy is needed for the process

Be careful with this type of question. You may believe that 'up a gradient' could be given for active transport. It is correct, but it's too close to the 'down a gradient' idea for facilitated diffusion.

Go for a completely different idea, as shown.

(c) What effect would the following temperatures have on the active transport of Mg^{2+} ions across a cell surface membrane of a plant cell? Assume the plant is a British native.   [4]

(i) 30°C

> It is likely that active transport would be efficient because the temperature would be ideal for the Mg^{2+} to bind with a carrier molecule.

(ii) 80°C

> process likely not to work;
> protein carrier denatured;
> Mg^{2+} would not be able to bind.

Practice examination questions

1

cell A   cell B   cell C

Cells A, B and C have been placed in different concentrations of salt solution.

(a) Explain each of the following in terms of water potential.

    (i)   Cell A did not change size at all.

    (ii) Cell B decreased in volume.

    (iii) Cell C became swollen and burst. [3]

(b) Which process is responsible for the changes to cells B and C? [1]

2 (a) Give one similarity between active transport and facilitated diffusion. [1]

  (b) Give one difference between active transport and facilitated diffusion. [1]

3 The diagram shows a section through a leaf.

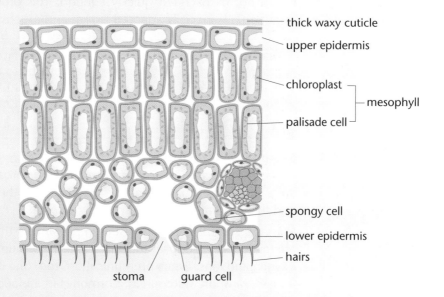

thick waxy cuticle
upper epidermis
chloroplast
mesophyll
palisade cell
spongy cell
lower epidermis
hairs
stoma   guard cell

Describe how the leaf is adapted for efficient photosynthesis. [6]

4 Describe and explain how the following are adapted to efficient gaseous exchange.

  (a) The alveoli of lungs. [4]

  (b) The gills of a fish. [4]

Chapter 5

Transport

The following topics are covered in this chapter:

- *Mass transport systems*
- *Heart: structure and function*
- *Blood vessels*

- *The transport of substances in the blood*
- *The transport of substances in a plant*

5.1 Mass transport systems

LEARNING SUMMARY

After studying this section you should be able to:

- *explain why most multicellular organisms need a mass transport system*
- *understand the importance of a high surface area to volume ratio*

Why do most multicellular organisms need a mass transport system?

AQA   2.7

The bigger an organism is, the lower its surface area to volume ratio. Substances needed by a large organism could not be supplied through its exposed external surface. Oxygen passing through an external surface would be rapidly used up before reaching the many layers of underlying cells. Similarly waste substances would not be excreted quickly enough. This problem has been solved, through evolution, by specially adapted tissues and organs.

> **KEY POINT**
>
> Leaves, roots, gills and lungs all have high surface area to volume properties so that supplies of substances vital to **all** the living cells are made available by these structures. Movement of substances to and from these structures is carried out by efficient **mass transport systems.**

This is a little like transport on trains where people travel together on the same train, in the same direction, at the same speed, but may get off at different places.

In a mass transport system, all the substances move in the same direction at the same speed. Across the range of multicellular organisms found in the living world are a number of mass transport systems, e.g. the mammalian circulatory system and the vascular system of a plant.

Mass transport systems are just as important for the rapid removal of waste as they are for supplies. Supplies include an immense number of substances, e.g. glucose, oxygen and ions. Even communication from one cell to another can take place via a mass transport system, e.g. hormones in a blood stream.

The greater the metabolic rate of an organism, the greater the demands on its mass transport system.

5.2 Heart: structure and function

After studying this section you should be able to:

- *recall the structure, cardiac cycle and electrical stimulation of a mammalian heart*

The mammalian heart

AQA ▶ 1.5

The heart consists of a range of tissues. The most important one is cardiac muscle. The cells have the ability to contract and relax through the complete life of the person, without ever becoming fatigued. Each cardiac muscle cell is **myogenic**. This means it has its own inherent rhythm. Below are diagrams of the heart and its position in the circulatory system.

Note that tricuspid and bicuspid valves are known as atrioventricular valves.

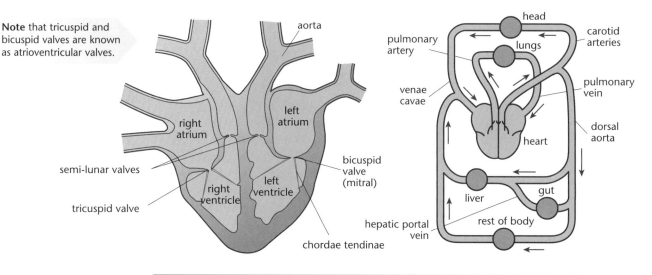

Structure

If blood moved in the wrong direction, then transport of important substances would be impeded.

The heart consists of four chambers, **right** and **left atria** above **right** and **left ventricles**. The functions of each part are as follows.

- The **right atrium** links to the **right ventricle** by the **tricuspid valve**. This valve prevents backflow of the blood into the atrium above, when the ventricle contracts.

- The **left atrium** links to the **left ventricle** by the **bicuspid valve (mitral valve)**. This also prevents backflow of the blood into the atrium above.

Check out these diagrams of a valve.

You can work out if a valve is open or closed in terms of pressure. Higher pressure above than below a semi-lunar valve closes it. Higher pressure below the semi-lunar valve than above, opens it.

- The **chordae tendonae** attach each ventricle to its **atrioventricular valve**. Contractions of the ventricles have a tendency to force these valves up into the atria. Backflow of blood would be dangerous, so the chordae tendonae hold each valve firmly to prevent this from occurring.

- Semi-lunar (pocket) valves are found in the blood vessels leaving the heart (pulmonary artery and aorta). They only allow exit of blood from the heart through these vessels following ventricular contractions. Elastic recoil of these arteries and relaxation of the ventricles closes each semi-lunar valve.

- Ventricles have thicker muscular walls than atria. When each atrium contracts it only needs to propel the blood a short distance into each ventricle.

- The left ventricle has even thicker muscular walls than the right ventricle. The left ventricle needs a more powerful contraction to propel blood to the systemic circulation (all of the body apart from the lungs). The right ventricle propels blood to the nearby lungs. The contraction does not need to be so powerful.

Cardiac cycle

Blood must continuously be moved around the body, collecting and supplying vital substances to cells as well as removing waste from them. The heart acts as a pump using a combination of **systole** (contractions) and **diastole** (relaxation) of the chambers. The cycle takes place in the following sequence.

Stage 1

Ventricular diastole, atrial systole
Both ventricles relax simultaneously. This results in lower pressure in each ventricle compared to each atrium above. The atrioventricular valves open partially. This is followed by the atria contracting which forces blood through the atrioventricular valves. It also closes the valves in the vena cava and pulmonary vein. This prevents backflow of blood.

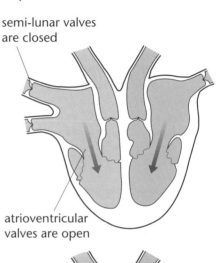

semi-lunar valves are closed

atrioventricular valves are open

Stage 2

Ventricular systole, atrial diastole
Both atria then relax. Both ventricles contract simultaneously. This results in higher pressure in the ventricles compared to the atria above. The difference in pressure closes each atrioventricular valve. This prevents backflow of blood into each atrium. Higher pressure in the ventricles compared to the aorta and pulmonary artery opens the semi-lunar valves and blood is ejected into these arteries. So blood flows through the systemic circulatory system via the aorta and vena cava and through the lungs via the pulmonary vessels.

Stage 3

Ventricular diastole, atrial diastole
Immediately following ventricular systole, both ventricles and atria relax for a short time. Higher pressure in the aorta and pulmonary artery than in the ventricles closes the semi-lunar valves. This prevents the backflow of blood. Higher pressure in the vena cava and pulmonary vein than in the atria results in the refilling of the atria.

The cycle is now complete – *GO BACK TO STAGE 1!*

The whole sequence above is **one** cardiac cycle or heartbeat and it takes less than one second. The number of heartbeats per minute varies to suit the activity of an organism. Vigorous exercise is accompanied by an increase in heart rate to allow faster collection, supply and removal of substances because of enhanced blood flow. Conversely during sleep, at minimum metabolic rate, heart rate is correspondingly low because of minimum requirements by the cells.

How is the heart rate controlled?

It has already been stated that the cardiac muscle cells have their own inherent rhythm. Even an individual cardiac muscle cell will contract and relax on a microscope slide under suitable conditions. An orchestra would not be able to play music in a coordinated way without a conductor. The cardiac muscle cells must be

---

Examination questions often test your knowledge of the opening and closing of valves. Always analyse the different pressures given in the question. A greater pressure behind a valve opens it. A greater pressure in front closes it.

Returning to Stage 1, the cycle begins again. The hormone adrenaline increases the heart rate still further. Even your examinations may increase your heart rate!

SAN

AVN

Purkinje tissue

The SAN is the natural pacemaker of the heart.

All of the Purkinje fibres together are known as the **Bundle of His**.

This is one of the examiners' favourite ways to test heart-related concepts. Look at the **peak** of the **ventricular contraction**. It coincides with the **trough** in the **ventricular volume**. This is not surprising, because as the ventricle contracts it empties! Use the data of higher pressure in one part and lower in another to explain:

(a) movement of blood from one area to another
(b) the closing of valves.

similarly coordinated, by a **pacemaker** area in the heart. Electrical stimulation from the brain can alter the activity of the pacemaker and therefore change the rate and strength of the heartbeat.

- The heart control centre in the brain is in the medulla oblongata.
- The sympathetic nerve stimulates an increase in heart rate.
- The vagus nerve stimulates a decrease in heart rate.
- These nerves link to a pacemaker structure in the wall of the right atrium, the **sinoatrial node (SAN)**.
- A wave of electrical excitation moves across both atria.
- They respond by contracting (the right one slightly before the left).
- The wave of electrical activity reaches a second pacemaker, the **atrioventricular node (AVN)**, which conducts the electrical activity through the **Purkinje fibres**.
- These Purkinje fibres pass through the septum of the heart deep into the walls of the left and right ventricles.
- The ventricle walls begin to contract from the apex (base) upwards.
- This ensures that blood is ejected efficiently from the ventricles.

Graphs to show the changes in pressure and volume during the cardiac cycle

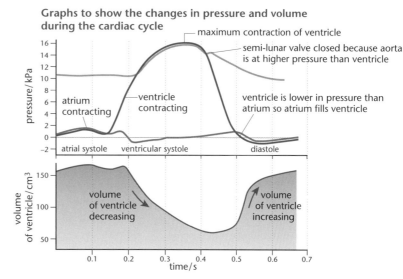

Cardiac output

The cardiac output is the volume of blood that is pumped out by one of the ventricles in a minute. It can be worked out using the formula:

Cardiac Output = Stroke Volume x Heart Rate
(volume of blood pumped out per beat by one ventricle)

Progress check

The medulla oblongata can increase the heart rate. The statements below include all of the events which take place, but in the wrong order. Write them out in the correct sequence.

A this ensures that blood is ejected efficiently from the ventricles

B the wave of electrical activity reaches the **atrioventricular node (AVN)** which conducts the electrical activity through the **Purkinje fibres**

C a wave of electrical excitation moves across both atria

D the sympathetic nerve conducts electrical impulses

E electrical impulses are received at the **sinoatrial node (SAN)**

F as a result the atria contract

G the ventricle walls begin to contract from the apex (base) upwards

D E C F B G A

5.3 Blood vessels

After studying this section you should be able to:

- describe the structure and functions of arteries, capillaries and veins
- understand the importance of valves in the return of blood to the heart
- understand the difference between plasma, tissue fluid and lymph

LEARNING SUMMARY

Arteries, veins and capillaries

AQA ▶ 2.7

The blood is transported to the tissues via the vessels. The main propulsion is by the ventricular contractions. Blood leaves the heart via the arteries, reaches the tissues via the capillaries, then returns to the heart via the veins. Each blood vessel has a space through which the blood passes; this is the **lumen**. The structure of the vessels is shown below.

Artery

Note that the pressure in the **arteries** is highest because:

(a) they are closest to the ventricles
(b) they contract forcefully themselves.

Capillaries are the next highest in pressure, the main factor being their resistance to blood flow.

Finally, the pressure of **veins** is the lowest because:

(a) they are furthest from the ventricles
(b) they have a low amount of muscle.

If given blood pressures of vessels, be ready to predict the correct direction of blood flow.

- It has a thick **tunica externa** which is an outer covering of tough collagen fibres.
- It has a **tunica media** which is a middle layer of **smooth muscle** and **elastic fibres**.
- It has a lining of **squamous endothelium** (very thin cells).
- It can contract using its **thick muscular layer**.

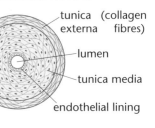

tunica externa (collagen fibres)
lumen
tunica media
endothelial lining

Capillary

- It is a very thin blood vessel, the endothelium is just one cell thick.
- Substances can exchange easily.
- It has such a high resistance to blood flow that blood is slowed down. This gives more time for efficient exchange of chemicals at the tissues.

endothelium
lumen

Vein

- It has a thin **tunica externa** which is an outer covering of tough collagen fibres.
- It has a very thin **tunica media** which is a middle layer of **smooth muscle** and **elastic fibres**.
- It has a lining of **squamous endothelium** (very thin cells).
- It is lined with semi-lunar valves which prevent the backflow of blood.

tunica externa (collagen fibres)
lumen
tunica media
endothelial lining

How do the veins return the blood to the heart?

Veins have a thin tunica media, so only mild contractions are possible. They return blood in an unexpected way. Every time the organism moves physically, blood is squeezed between skeletal muscles and forced along the vein.

direction of blood flow

semi-lunar valve

> Blood must travel towards the heart because of the direction of the semi-lunar valves. Any attempt at backflow and the semi-lunar valves shut tightly!

KEY POINT

Capillary network

Every living cell needs to be close to a **capillary**. The arteries transport blood from the heart but before entry into the capillaries it needs to pass through smaller vessels called **arterioles**. Many arterioles contain a ring of muscle known as a **pre-capillary sphincter**. When this is contracted the constriction shuts off blood flow to the capillaries, but when it is dilated, blood passes through. Some capillary networks have a shunt vessel. When the sphincter is constricted blood is diverted along the shunt vessel so the capillary network is by-passed. After the capillary network has permeated an organ the capillaries link into a **venule** which joins a **vein**.

In the skin the superficial capillaries have the arteriole/shunt vessel/venule arrangement as shown opposite. When the arteriole is dilated (**vasodilation**) more heat can be lost from the skin. When the arteriole is constricted (**vasoconstriction**) the blood cannot enter the capillary network so is diverted to the core of the body. Less heat is lost from the skin.

Although the pressure of the blood in the capillaries is lower than in the arteries or arterioles, there is still enough pressure to force out some of the liquid part of the blood. The liquid part of the blood is called **plasma** and when it is forced out of the capillaries it is called **tissue fluid**.

This tissue fluid bathes the cells, supplying them with nutrients and taking up waste products. At the venous end of the capillary bed, most of this tissue fluid is reabsorbed back into the capillaries.

Lymphatic system

There is a network of vessels other than the blood system. They are the **lymphatic vessels**. They collect any tissue fluid that is not reabsorbed back into the capillaries.

The lymph vessels have valves to ensure transport is in one direction. Along some parts of the lymphatic system are lymph nodes. These are swellings lined with white blood cells (macrophages and lymphocyte cells). The lymph fluid is finally emptied back into the blood near the heart.

Progress check

The diagram shows the structure of a blood vessel.

(a) (i) Which type of vessel, artery, vein or capillary is shown? Give a reason for your choice.

   (ii) What is the function of the tunica media?

(b) The pressure values 30 kPa, 10 kPa and 5 kPa correspond to the different types of vessel. Give the correct value for each vessel so that blood flows around the body.

(a) (i) artery; the vessel has a thick tunica externa
(ii) contracts to help transport blood.
(b) artery, 30 kPa, capillary, 10 kPa and vein, 5 kPa.

5.4 The transport of substances in the blood

After studying this section you should be able to:

- *describe the transport of oxygen in the blood and explain how oxygen is released at the tissues*

LEARNING SUMMARY

How is oxygen transported?

AQA 2.4

Oxygen is absorbed in the lungs from fresh air which has been breathed in. Red blood cells (**erythrocytes**) contain the protein **haemoglobin** which can reversibly combine with oxygen. In the lungs, where the concentration of oxygen is high, haemoglobin will take up oxygen and form oxyhaemoglobin. In the tissues where the oxygen concentration is low, the oxyhaemoglobin will dissociate and release the oxygen. This is shown by the graph below which is called the oxygen dissociation curve.

Features of the dissociation curve

- At high partial pressures of oxygen, haemoglobin has a high affinity (attraction) for oxygen and is highly saturated.
- At low partial pressures, the affinity is lower and the oxyhaemoglobin dissociates and is less saturated.
- The curve is sigmoid or 'S' shaped. This means that the curve is steep and a small change in partial pressure causes a massive loading or unloading of oxygen.

Changes to the dissociation curve

Different factors can cause changes to the dissociation curve:

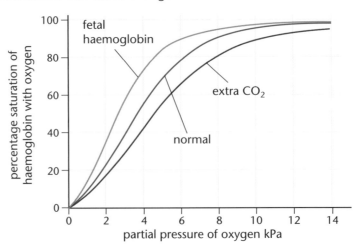

- The greater the amount of carbon dioxide at the tissues, the more the dissociation curve is moved to the right, and the more oxygen is 'off-loaded' to the tissues. This is called the Bohr shift. The carbon dioxide lowers the affinity of the haemoglobin for oxygen.
- Fetal haemoglobin has a greater affinity for oxygen than adult haemoglobin. This allows the fetus to take oxygen from the mother's haemoglobin.

Myoglobin

Myoglobin only has one polypeptide chain compared to four found in each molecule of haemoglobin. Myoglobin is often found in diving animals or animals that live in anaerobic mud, e.g. lugworms.

There is another substance that can act as a respiratory pigment. This is **myoglobin**. It does not travel in the blood but is found in muscle. It has a greater affinity for oxygen than haemoglobin and so only releases oxygen at very low partial pressures. It acts as a store of oxygen trying to prevent anaerobic respiration from occurring.

5.5 The transport of substances in a plant

After studying this section you should be able to:

- *recall the structure of a root and understand how water and ions are absorbed*
- *recall the structure of xylem and phloem and explain the processes by which they transport essential chemicals*
- *understand how plants lose water and how this loss can be measured*
- *recall the adaptations of xerophytes*

LEARNING SUMMARY

Root structure and functions

AQA   2.7

The roots of a green plant need to exchange substances with the soil environment. The piliferous zone just behind a root tip has many root hairs which have a high surface area to volume ratio.

- Root hairs are used for absorption of water and mineral ions and the excretion of carbon dioxide.

Note that the root hairs also absorb oxygen from the air to aid aerobic respiration. The high surface area to volume ratio certainly helps!

- They have a cell membrane with a high surface area to volume ratio to efficiently absorb water, mineral ions and oxygen, and excrete carbon dioxide.
- They project out into the soil particles which are surrounded by soil water at high water potential compared with the low water potential of the contents of the root hairs.

Remember that water moves from a higher water potential to a more negative water potential.

- They have a cell membrane which is partially permeable to allow water absorption by osmosis (see page 50).
- As they absorb more water by osmosis, the cell sap becomes more dilute compared with neighbouring cells. Water therefore moves to these adjacent cells which become more diluted themselves, so osmosis continues across the cortex.

Remember that active transport needs energy, so mitochondria will be close to the carrier molecules on the membranes.

- They have carrier proteins in the cell membranes to allow mineral ions to be absorbed by active transport.

Passage of water into the vascular system

Once absorbed by osmosis, water needs to pass to the xylem vessels in order to move up the plant. First it must move across the cortex of the root and through

Note the different theories for water transport across the width of the root.

the endodermis before entering the xylem. The mechanism of passage is not known but there are three theories:

- **apoplast** route, where the water is considered to pass between the cells
- **symplast** route, where the water is considered to pass via the cytoplasm of the cells via **plasmodesmata** (cytoplasmic strands connecting one cell to another)
- **vacuolar** route, where the water is considered to pass through the tonoplast then through the sap vacuole of each cell.

movement of water across the cortex

Casparian strip

Water moves across the cortex and needs to pass through the endodermal cells before entering the xylem vessels of the vascular system. Each cell of the endodermis has a waterproof band around it, just like a ribbon around a box. This means that water must pass through the cell in some way, rather than around the outside. If water moves by the apoplast route up to this point, then it must now move into the symplast or vacuolar pathways.

Casparian strip of the endodermal cells

Once the water has passed through the endodermis and navigated the pericycle then it must pass into the xylem for upward movement to the leaves and to the tissues.

How does water move up the vascular system to the leaves?

Water moves into the **xylem** vessels in the vascular system in the centre of the root; it enters via **bordered pits**. The xylem is internally lined with **lignin**. This substance is waterproof and it also gives great strength to the xylem vessels, which are tube shaped. Much of the strength of a plant comes from cells toughened by lignin. A Giant Redwood tree is many metres high but water is still able to reach all the cells. Water moves up the xylem for the following reasons.

Remember that the xylem is part of the mass flow system ensuring that all cells receive their requirements.

- **Root pressure** gives an initial upward force to water in the xylem vessels. This can be shown by cutting off a shoot near soil level. Some sap will pour vertically out of the xylem of the remaining exposed xylem.

- Water moves up the xylem by **capillarity** which is the upward movement of a fluid in a narrow bore tube – xylem has very narrow vessels.

The factors in the list are known as the cohesion-tension theory and explain how water moves up the xylem.

- Capillarity occurs because the water molecules have an attraction for each other (cohesion) so when one water molecule moves others move with it.

- Capillarity has another component – the fact that the water molecules are attracted to the sides of the vessels pulls the water upwards (adhesion).

- Transpiration causes a very negative water potential in the mesophyll of the leaves. Water in the xylem is of higher water potential and so moves up the xylem.

Xylem vessels die at the end of their maturation phase. The lignin produced inside the cells finally results in death. The young xylem cells end to end, finally produce a long tube-like structure (vessel) through which water passes. Xylem can still transport water after the death of the plant.

Mineral ions are also transported in the xylem.

How is water lost from a leaf?

AQA ▶ 2.7

Water moves up the xylem and into the mesophyll of a leaf. The process by which water is lost from any region of a plant is **transpiration**. Water can be lost from areas such as a stem, but most water is lost by **evaporation** through the **stomata**. Each stoma is a pore which can be open or closed and is bordered at either side by a guard cell. The diagrams show an open stoma and a closed stoma.

thick wall
thin wall

stoma open   stoma closed

KEY POINT
Transpiration from a leaf takes place as follows:
• the air spaces in the mesophyll become **saturated** with water vapour (**higher water potential**)
• the air outside the leaf may be of **lower humidity** (more negative water potential)
• this causes water molecules to **diffuse** from the mesophyll of the leaf to the outside.

Some water can escape through the cell junctions and membranes. This is known as **cuticular transpiration**. In the dark all stomata are closed. Even so, there is still water loss by cuticular transpiration.

How do the guard cells open and close?

In the presence of light:

- K$^+$ ions are actively transported into the guard cells from adjacent cells
- malate is produced from starch
- K$^+$ ions and malate accumulate in the guard cells
- this causes an influx of water molecules
- the cell wall of each guard cell is thin in one part and thick in another
- the increase in hydrostatic pressure leads to the opening of the stomata.

Closing of the stomata is the reverse of this process. Under different conditions the stomata can be partially open. The rate of transpiration can increase in warm, dry conditions or decrease at the opposite extreme.

Measuring the rate of transpiration

This is done indirectly by using a potometer. This instrument works on the following principle: for every molecule of water lost by transpiration, one is taken up by the shoot.

Potometer

Note that a living shoot may be photosynthesising whilst attached to the instrument. Only a minute amount of water would be used in this process. The instrument gives an accurate measure of transpiration.

The potometer is used as follows:

- a shoot is cut and the end is quickly put in water to prevent an air lock in the xylem
- the potometer is filled under water so that the capillary tube is full
- all air bubbles are removed from the water
- the shoot is put into the rubber seal
- the valve is changed to allow water uptake
- the amount of water taken up by the shoot per unit time is measured
- the shoot can be tested under various conditions.

Xerophytes

AQA   2.7

These are plants which have **special adaptations** to survive in drying, environmental conditions where many plants would become desiccated and die. The plants survive well because of a combination of the following features:

- thick cuticle to reduce evaporation
- reduced number of stomata
- smaller and fewer leaves to reduce surface area
- hairs on plant to reduce air turbulence
- protected stomata to prevent wind access
- aerodynamic shape to prevent full force of wind
- deep root network to absorb maximum water
- some store of water in modified structures, e.g. the stem of a cactus.

> In an exam you may be given a photomicrograph of a xerophytic plant which you have not seen before. Look for *some* of the features covered in the bullet points opposite.

A cactus is a good example of a xerophyte. It makes excellent use of what little water there is available, and holds on to what it does manage to absorb really well. Its cuticle and epidermis are so thick that metabolic water released from the cells during night time respiration is retained for photosynthesis during the day. Nothing is wasted!

Progress check

Water is absorbed into a plant by the root hairs.

(a) The water potential of the root hair cells is more negative than in the soil water. Is this statement true or false?

(b) Describe:

(i) the apoplast route across the cortex

(ii) the symplast route across the cortex.

(a) true

(b) (i) water is considered to pass on the outside of the cell membrane

(ii) water passes through the cytoplasm of the cells through plasmodesmata.

Sample question and model answer

(a) The graph shows the oxygen dissociation curve for human haemoglobin.

Use the information in the graph to help you answer the following questions.

(i) What is the advantage of haemoglobin as a respiratory pigment when oxygen in the air is at the low partial pressure 6 kPa? [1]

Even at a low partial pressure a lot of oxygen (70%) is taken up by the haemoglobin of a red blood cell.

(ii) Explain the effect on the oxygen dissociation curve of a high partial pressure of carbon dioxide at a muscle. [2]

The curve is moved to the right and down so that oxygen is released.

(iii) Fetal haemoglobin has a greater affinity for oxygen than maternal haemoglobin. Draw a curve on the graph to show the oxygen dissociation curve for fetal haemoglobin. [1]

See graph opposite.

The fact that haemoglobin is able to carry oxygen is important. However, it is just as important that the oxygen is off-loaded at tissues needing it. This is only possible because carbon dioxide is found at the tissues.

(b) The diagram shows **one** stage in the cardiac cycle.

Always look for the valves. If the heart valve is open then the chamber behind it is contracting.

(i) Which stage of the cardiac cycle is shown in the diagram? Give **two** reasons for your answer. [3]

atrial systole
the atrioventricular valves are open/blood flows through the atrioventricular valves,
semi-lunar valves are closed.

(ii) Write an X in one chamber to show the position of the atrioventricular node (AVN). [1]

(iii) How does the AVN stimulate the contraction of the ventricles? [1]

Passes electrical impulses to Purkinje tissue/Bundle of His.

Practice examination questions

1   The diagram shows a capillary bed in the upper part of the skin. The arteriole is constricted.

Use the information in the diagram and your own knowledge to answer the questions below.

(a) As a result of arteriole constriction, to where would the blood flow?   [1]

(b) Explain how this would help maintain the body temperature.   [4]

2   The table shows data about a person's heart before and after a training programme.

	Before training	After training
heart stroke rate	90 ml	120 ml
heart rate at rest	75 bpm	60 bpm
maximum heart rate	170 bpm	190 bpm

(a) Over a five minute period at rest before training, the cardiac output of the person was 33.75 litres.

How much blood would leave the heart, during the same time, whilst the person was at rest, after training?   [2]

(b) After training, the maximum heart rate increased by 20 bpm. Explain the advantage of this increase to an athlete.   [4]

(c) After training there are other changes in the body.

Explain:

(i)   **two** changes which would improve the efficiency of the respiratory system.   [2]

(ii) the effect of training on the muscles.   [2]

Practice examination questions (continued)

3 The diagram shows a freshly cut, leafy shoot attached to a potometer. This was used to measure the amount of water taken up by the shoot under different conditions.

(a) What assumption must be made when using this apparatus to measure the rate of transpiration? [1]

(b) An air-lock can occur in the shoot which prevents water uptake.

(i) In which plant tissue could an air-lock occur? [1]

(ii) Describe the practical details by which a student could make sure that there was no air-lock in the shoot. [2]

(c) The radius of the capillary tube of the potometer was 1 mm. When a Sumach leaf was measured the air bubble moved 32 mm in one minute. Calculate the volume of water in mm^3 which would be taken up by the leaf in one hour under the same environmental conditions. [3]

4 *Agave americana* is a xerophytic plant which grows in the deserts of Mexico.

Agave americana

Suggest **three** ways in which the plant is adapted to survive periods of very low rainfall. [3]

5 The diagram shows nerves linking the medulla oblongata with the heart.

(a) Name part X. [1]

(b) What effect do the following have on the heart:

(i) vagus nerve

(ii) sympathetic nerve

(iii) adrenaline? [3]

Chapter 6

Genes and cell division

The following topics are covered in this chapter:

- DNA and the genetic code
- Cell division

6.1 DNA and the genetic code

After studying this section you should be able to:

- recall the structure of DNA
- outline the role of DNA in the synthesis of protein
- use organic base codes of DNA to identify amino acid sequences

LEARNING SUMMARY

Deoxyribonucleic acid (DNA) and chromosome structure

AQA   2.2, 2.3

Each chromosome in a nucleus consists of a series of genes. A gene is a section of a chemical called **DNA** and each gene controls the production of a polypeptide important to the life of an organism. **Deoxyribonucleic acid (DNA)** is made up of a number of **nucleotides** joined together in a double helix shape.

> You need to be aware that many nucleotides join together to form the polymer, DNA.

Each nucleotide consists of a phosphate group, a molecule of deoxyribose sugar and an organic base. Phosphate and pentose sugar units link to form the backbone of the DNA. Repeated linking of the monomer nucleotides forms the polynucleotide chains of DNA.

> Each strand of DNA is said to be complementary to the other. **Examination tip:** be ready to identify one strand when given the matching complementary strand.

Each DNA molecule is made up of two polynucleotide chains. The two chains are held together by hydrogen bonds between the bases.

The organic base of each nucleotide can be any one of **adenine**, **thymine**, **cytosine** or **guanine**. Adenine forms hydrogen bonds with thymine and cytosine with guanine.

The two chains then twist up to form a double helix.

Why does the DNA of one organism differ from the DNA of another?

Differences in the DNA of organisms such as humans and houseflies lie in the **different sequences** of the organic bases. Each sequence of bases is a code to make a protein, usually vital to the life of an organism.

sugar phosphate

organic bases

A single nucleotide

phosphate

pentose sugar

organic base

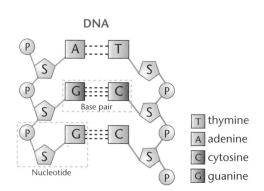

DNA

Base pair

Nucleotide

T	thymine
A	adenine
C	cytosine
P	guanine
G	guanine

How does DNA control protein production?

AQA 2.2

Each different polypeptide is made of a specific order of amino acids and so DNA must code for this order. The length of DNA that codes for one polypeptide is called a gene.

> **KEY POINT**
>
> Each three adjacent bases code for one amino acid in the polypeptide. The genetic code is therefore a **triplet** code.

Use the key to identify the amino acids in the table opposite.

Amino acid	Abbreviation
alanine	Ala
arginine	Arg
asparagine	Asn
aspartic acid	Asp
cysteine	Cys
glutamine	Gln
glutamic acid	Glu
glycine	Gly
histidine	His
isoleucine	Iso
leucine	Leu
lysine	Lys
methionine	Met
phenylalanine	Phe
proline	Pro
serine	Ser
threonine	Thr
tryptophan	Trp
tyrosine	Tyr
valine	Val

The table below shows all the triplet sequences of organic bases found along DNA strands and the coding function of each.

Genetic code functions of DNA

	Second organic base								Third organic base
	A		G		T		C		
A	AAA	Phe	AGA		ATA	Tyr	ACA	Cys	A
	AAG		AGG	Ser	ATG		ACG		G
	AAT	Leu	AGT		ATT	stop	ACT	stop	T
	AAC		AGC		ATC	stop	ACC	Trp	C
G	GAA		GGA		GTA	His	GCA		A
	GAG	Leu	GGG	Pro	GTG		GCG	Arg	G
	GAT		GGT		GTT	Gln	GCT		T
	GAC		GGC		GTC		GCC		C
T	TAA		TGA		TTA	Asn	TCA	Ser	A
	TAG	Ile	TGG	Thr	TTG		TCG		G
	TAT		TGT		TTT	Lys	TCT	Arg	T
	TAC	Met	TGC		TTC		TCC		C
C	CAA		CGA		CTA	Asp	CCA		A
	CAG	Val	CGG	Ala	CTG		CCG	Gly	G
	CAT		CGT		CTT	Glu	C CT		T
	CAC		CGC		CTC		CCC		C

Each triplet code is **non-overlapping**. This means that each triplet of three bases is a code, then the next three, and so on along the DNA.

- AAA codes for the amino acid phenylalanine
- GAG codes for the amino acid leucine
- GAC codes for the amino acid leucine

There are more triplet codes than there are amino acids. This is known as the **degenerate code**, because an amino acid such as leucine can be coded for by up to six different codes. Some triplets do not code for amino acids but mark the beginning or end of polypeptides. They are **stop or start** triplets.

Introns

In 1977, it was discovered that many genes in eukaryotic organisms were 'split genes'. Mixed in with the DNA that codes for the protein are lengths of DNA that do not code for anything. These lengths are called **introns**. There are also areas of bases between genes that do not seem to code for any proteins.

Some of these non-coding lengths of DNA have unique repeat patterns that are used in genetic finger-printing.

Progress check

(a) Name the parts of a nucleotide.
(b) (i) By which bonds do the two strands of DNA link together?
    (ii) How would these bonds be broken in the laboratory to produce single strands of the DNA?

(a) pentose sugar, phosphate and organic base. The organic base may be thymine, adenine, cytosine or guanine
(b) (i) hydrogen bonds (ii) heat

6.2 Cell division

After studying this section you should be able to:

- describe and explain the semi-conservative replication of DNA
- understand that DNA must replicate before cell division can begin
- recall the purpose of mitosis and meiosis
- recognise each stage of the cell cycle and cell division by mitosis

LEARNING SUMMARY

How do cells prepare for division?

AQA 2.5

Before cells divide they must first make an exact copy of their DNA by using a supply of organic bases, pentose sugar molecules and phosphates. The method by which DNA is copied is called **semi-conservative replication**. The diagram (right) shows this taking place.

Remember that as the DNA unwinds each single strand is a **complement** to the other. This means that each has the **matching** series of organic bases.

- The DNA begins to unwind under the influence of the enzyme DNA helicase.
- Hydrogen bonds between the two chains then break and the two strands separate.
- Each complementary strand then acts as a template to build its opposite strand from free nucleotides.
- The enzyme DNA polymerase joins the nucleotides together. This process results in the production of two identical copies of double-stranded DNA.

Exam questions are often based on experimental data. Apply your knowledge of principles learned during the course and this will be helpful.

Evidence for the semi-conservative replication of DNA

The first real evidence came from the results of an experiment carried out by two researchers, Meselson and Stahl.

Bacteria were cultured with a heavy isotope of nitrogen located in the organic bases of their DNA.

The bacteria were then supplied with bases containing the normal light nitrogen atoms. They replicated their DNA using these bases. Their population increased.

Each molecule of DNA of the next generation had one strand containing heavy nitrogen and one strand containing light nitrogen. The mass of the DNA was therefore midway between the original heavy form and normal light DNA.

Semi-conservative replication

both strands are heavy → one strand is heavy and the other is light

heavy DNA

light DNA

Semi-conservative therefore means that as DNA splits into its two single strands, each of the new strands is made of newly acquired bases. The other strand, part of the original DNA, remains.

Make sure that you can predict what would happen if the bacteria replicated again.

Types of cell division

AQA ▷ 2.5

Cells divide for the purposes of growth, repair and reproduction. Not all cells can divide but there are two ways in which division may occur: **mitosis** and **meiosis**.

In most organisms, the chromosomes in each cell can be arranged in pairs. Each cell therefore has two copies of each gene.

When the chromosomes are in pairs, the cell is said to be **diploid** and the pairs are called homologous pairs.

Mitosis produces genetically identical copies of cells with the same number of chromosomes for growth and repair.

Be careful of the spelling of mitosis and meiosis; examiners may penalise you if they are spelt incorrectly.

Meiosis produces cells that have half the number of chromosomes, one from each pair. These cells are **haploid** and are used as gametes.

Meiosis introduces variation because the pairs of chromosomes can be split up in many different ways. This is called independent assortment.

Genetic material can also be exchanged between homologous chromosomes. This is called **crossing over**.

The cell cycle and mitosis

AQA ▷ 2.5

The length of time between a cell being formed and it dividing is called the **cell cycle**. This can be divided up into a number of different phases:

- G1 phase – the cell grows making new proteins and more organelles
- S phase – the DNA of the chromosomes is replicated by semi-conservative replication
- G2 – more organelles are made and a spindle forms
- M phase – this is mitosis involving the separating of the genetic material into two nuclei
- C phase – cytokinesis, where the cell divides into two.

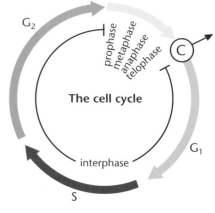

G_2 — prophase metaphase anaphase telophase — C

The cell cycle

G_1

interphase

S

Remember that DNA replication takes place before cell division in **interphase** (see page 76). This is not an integral phase of mitosis or meiosis.

The length of the cell cycle varies between different types of cells. Some cells never divide once formed but cells of the bone marrow divide about every eight hours. Interphase usually takes up 95% or more of the whole cell cycle.

Be ready to analyse photomicrographs of all phases of mitosis. If you can spot 10 pairs of chromosomes at the end of telophase, then this is the original diploid number of the parent cell.

Interphase is the period of time between cell divisions. It is made up of G1, S and G2 phases. Once cells start mitosis, they all go through a similar sequence of events. This is shown in the diagrams.

1 Prophase

Each chromosome forms two chromatids joined by a centromere. Two centrioles begin to move forming a spindle.

chromatid

centriole

2 Metaphase

The chromatids, still joined by a centromere move to the middle of the cell. Each of the two chromatids has identical DNA to the other.

spindle

3 Anaphase

The spindle fibres join to the centromeres. The spindle fibres shorten and the centromeres split. The separated chromatids are now chromosomes.

4 Telophase

Identical chromosomes move to each pole. The nuclear membrane re-forms. The cell membrane narrows at the middle and two daughter cells are formed.

Practice examination questions

1 The diagram below shows a stage in the process of mitosis.

(a) Give the stage of mitosis shown. [1]

(b) How many chromosomes would there be in the daughter cells? [2]

2 The table below shows the relative organic base proportions found in human, sheep, salmon and wheat DNA.

Organism	Proportion of organic bases in DNA (%)			
	Adenine	Guanine	Thymine	Cytosine
human	30.9	19.9	29.4	19.8
sheep	29.3	21.4	28.3	21.0
salmon	29.7	20.8	29.1	20.4
wheat	27.3	22.7	27.1	22.8

(a) Refer to the proportion of organic bases in salmon DNA to explain the association between specific bases. [2]

(b) Suggest a reason for the small difference in proportion of the organic bases adenine and thymine in sheep. [1]

(c) All species possess adenine, guanine, thymine and cytosine in their DNA. Account for the fact that each species is different. [2]

Chapter 7

Classification and biodiversity

The following topics are covered in this chapter:

- *Classification*
- *Biodiversity*

7.1 Classification

After studying this section you should be able to:

- *understand why and how organisms are classified*
- *describe the binomial system of naming organisms*
- *describe some of the more recent techniques used to classify organisms*

LEARNING SUMMARY

Classifying and naming

AQA   2.8

It is estimated that there are more than 10 million different kinds of living organisms alive on Earth. The study of the range of different organisms is called **systematics**. For centuries, scientists have tried to classify organisms into groups and give them names.

> The system of classifying organisms into groups is called **taxonomy** and the system of naming is called **nomenclature**.
>
> **KEY POINT**

Organisms are put into groups based on the largest number of common characteristics. Firstly, they are put into one of five large groups called kingdoms.

The table shows the characteristics of organisms in the five kingdoms:

Try this!

Pretty **P**olly **F**inds **P**arrots **A**ttractive.

It will help you to remember the kingdoms!

You may be tested on your knowledge of characteristics across the kingdoms.

Information will usually be given for any subgroup tasks.

The table includes some of the main features of each kingdom.

Prokaryotae and Protoctista tend to give more of a challenge than some of the other kingdoms, and are examined more often.

		→ *increase in complexity*		
Prokaryotae	*Protoctista*	*Fungi*	*Plantae*	*Animalia*
very simple cells with few organelles	unicellular cells with membrane-bound organelles	heterotrophic nutrition	multicellular organisms which are photosynthetic	multicellular organisms which are heterotrophic
no membrane-bound organelles	some are photo-synthetic, but many have heterotrophic nutrition	some sapro-trophic, some parasitic	cells have cellulose cell wall, sap vacuole and chloroplasts	no cell walls no sap vacuoles
if there are flagellae, then not 9 + 2 system of microtubules	reproduction usually involves fission	consists of thread-like hyphae, chitin cell walls	reproduce by seeds, or by spores, some sexual, some asexual.	
DNA in strands, no true nucleus		many nuclei in hyphae, not in one per cell organisation		
e.g. bacteria and cyanobacteria	e.g. algae and protozoa	reproduction involves the production of spores		

Each kingdom can be subdivided into a number of progressively smaller groups. Ultimately, this leads to an individual type of organism, a **species**.

The seven groups can be difficult to remember. Try the easy way!

King   Penguins
Climb   Over   Frozen
Grassy   Slopes

The first letter of each word will help you remember. It is a mnemonic which is an excellent strategy to aid recall.

The hierarchy of the groups is shown below.

	Example 1	*Example 2*
Kingdom	Animalia	Animalia
Phylum	Chordata	Arthropoda
Class	Mammalia	Insecta
Order	Primates	Lepidoptera
Family	Hominidae	Pieridae
Genus	Homo	Pieris
Species	sapiens	brassica

What is a species?

The smallest classification group is the species.

Remember, similar organisms such as horses and donkeys can mate to produce mules, but mules are infertile.

> Organisms are members of the same species if they can breed together to produce **fertile offspring**.
>
> **KEY POINT**

Naming organisms

Most organisms are known by common names. We use these names all the time. The problem is that an organism may be known by different names or sometimes different organisms can have the same common name. A scientist called Linnaeus introduced a scientific naming system so that each species could have a unique scientific name. This system is called the **binomial system**.

The name consists of two parts, the name of the genus written with a capital letter and the name of the species, written in lower case.

Notice that binomial names are always typed in italics or underlined if handwritten.

Panthera leo

Panthera tigris

Modern classification techniques

AQA ▶ 2.8

For centuries, organisms were classified according to observable physical features. This might be the structure of an animal's skull or the shape of a plant's leaves. Scientists can now use a range of different techniques.

- Microscopic structure – modern electron microscopes have shown that bacteria and other single-celled organisms such as amoeba have completely different cell structure and so they are now put into different kingdoms: *Prokaryotae* and *Protoctista*.

- Genetic differences – it is now possible to compare the DNA base sequences of different organisms. This can be done by a process called **DNA hybridisation**. Short, single-stranded sections of DNA are produced from one species and the extent to which they bind with DNA from another species is measured.
- Biochemical differences – the occurrence of different biochemical molecules is often a good indicator of relationships. It is also possible to work out and compare the amino acid sequence of common proteins.
- Immunological evidence – antibodies against human proteins can be made and then tested against proteins from other animals. The more effective the antibodies are, then the closer the evolutionary relationship is between humans and the other animal.

Immunological results give the following % similarities to man:

chimpanzee 97%,
gibbon 92%,
lemur 37%,
pig 8%

The aim of modern classification systems is to use a range of techniques to produce a system that classifies organisms based on their **evolutionary relationships**.

> **KEY POINT**
>
> A system based on evolutionary relationships is called a **phylogenetic system**.

7.2 Biodiversity

LEARNING SUMMARY

After studying this section you should be able to:

- *understand what is meant by the term biodiversity*
- *explain why there is so much biodiversity on Earth*

What is biodiversity?

AQA 2.11

Biodiversity is a measure of the variation between different living organisms. It could be:

- **genetic diversity** – the differences between the genes in a species
- **species diversity** – the number of different species in a community
- **ecosystem or habitat diversity** – the variety of different areas where organisms can live.

A pond is a habitat containing populations such as stickleback and pondweed. All the organisms make up the community and the organisms, the water, the mud, etc. are an ecosystem.

A few definitions

A **habitat**: an area in which an organism lives.
A **population**: all the organisms of one species living in a habitat.
A **community**: all the different species (different populations) living in a habitat.
An **ecosystem**: all the living organisms (biotic) and inorganic parts (abiotic) of a habitat.
A **niche** is where an organism lives and its role in that ecosystem.

Estimating populations

Often a full count of organisms in an ecosystem is not possible because of the size of the ecosystem. A **sampling technique** is used which requires a **quadrat**. This is a small area enclosed by wire or wood, around 0.25 m². When placed in the ecosystem the organisms inside the area can be counted, as well as the **abiotic factors** which influence their distribution. Ecologists use units to measure organisms within the quadrats. Frequency (f) is an indication of the presence of an organism in a quadrat area. This gives no measure of numbers. However the usual unit is that of density, the numbers of the organism per unit area. Sometimes percentage cover is used, an indication of how much of the quadrat area is occupied.

Point quadrat

string or wire

count organisms only at the intersections

Consider a survey of two species *Taraxacum officinale* (dandelion) and *Plantago major* (Great plantain) of the lawn habitat shown below.

A simplified results table

Quadrat no.	Dandelion
1	2
2	12
3	15
4	3
5	4
6	8
7	7
8	10
9	9
10	15

mean = 8.5 per quadrat
$Dm^{-2} = 34$

Lawn habitat

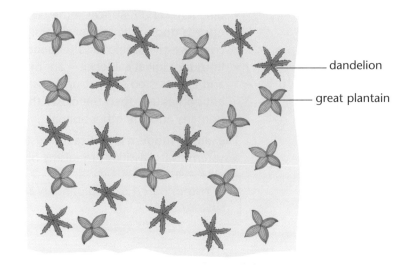

dandelion

great plantain

It is important to use a suitable technique when surveying with quadrats. When you observe a habitat which appears **homogeneous** or **uniform**, like a field yellow with buttercups, then you should use **random quadrat placement**. The area should be gridded, numbers given to each sector of the grid, then a random number generator used. The probability of the numbers in one quadrat representing a field would be very low. In practice the **mean** numbers from large numbers of quadrats do represent the true numbers in a habitat.

The index of biodiversity

This is used as a measure of the range and numbers of species in an area.

$$\text{index } d = \frac{N(N-1)}{\Sigma\, n(n-1)}$$

N = total no. of all individuals of all species in the area
n = total no. of individuals of one species in an area
Σ = the sum of

Consider this example of animals in a small pond

In another pond there were:

crested newt	45
stickleback	4
leech	18
great pond snail	10

d = 2.6

Look at both indices. 6.05 is an indicator of greater diversity. The higher number indicates greater diversity.

crested newt	8
stickleback	20
leech	15
great pond snail	20
dragonfly larva	2
stonefly larva	10
water boatman	6
caddisfly larva	30
N =	111

$$d = \frac{111 \times 110}{(8 \times 7) + (20 \times 19) + (15 \times 14) + (20 \times 19) + (2 \times 1) + (10 \times 9) + (6 \times 5) + (30 \times 29)}$$

$$d = \frac{12\,210}{2018} \quad \text{so } d = 6.05$$

Sources of biodiversity

AQA 2.1

What is variation?

Species throughout the biosphere differ from each other.

Variation describes the differences which exist in organisms throughout the biosphere. This variation consists of differences **between** species as well as differences **within** the same species. An organism's characteristics (phenotype) is a combination of the action of an organism's genes (genotype) and the influence of the environment.

genotype + environment = phenotype

Therefore variation can depend on environmental factors and genetic diversity. The variation that is produced can be either continuous or discontinuous.

Continuous variation

This is shown when there is a range of **small incremental differences** in a feature of organisms in a population. An example of this is height in humans. If the height of each pupil in a school is measured then from the shortest pupil to the tallest, there are very small differences across the distribution. This is shown by the graph below which shows smooth changes in height across a population. This type of variation is shown when features are controlled by **polygenic inheritance**. A number of genes **interact** to produce the expressed feature.

Remember that a species shows continuous variation when there are small incremental differences, e.g. height of people in a town. Beginning with the smallest and ending with the tallest there would probably be at least one person at each height, at 1cm increments. A smooth gradation of differences!

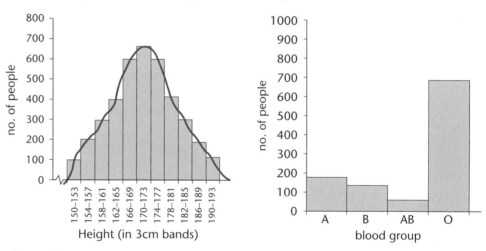

Discontinuous variation

This is shown when a characteristic is expressed in discrete categories. Humans have four discrete blood groups, A, B, AB or O. There are no intermediates, the differences are clear cut!

Mean and standard deviation

The variation shown by most populations can be described graphically. The variable usually forms a bell-shaped curve called a **normal distribution**.

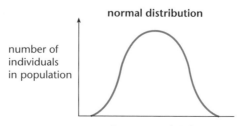

In a normal distribution, 68% of the values are within one standard deviation either side of the mean and 95% are within two.

The **mean** is the value at the peak of the curve. The **standard deviation** describes how the values are spread out. The higher the standard deviation, then the more variation there is.

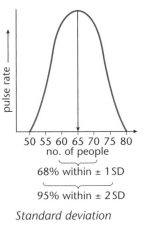

Standard deviation

The influence of the environment

The many different habitats on Earth provide different conditions for organisms to live in. There may be variations in a number of different factors that can affect organisms:

- **climatic** factors, e.g. temperature, water availability, light
- **edaphic** (soil) factors, e.g. pH, minerals and oxygen content
- **biotic** factors, e.g. competition.

Genetic diversity

Although organisms of the same species have the same genes, they show genetic variation because they have different alleles. These alleles are produced by **mutation**.

Sometimes genetic variation is reduced because factors can limit the variety of alleles. Some of these factors are manmade and others are naturally occurring.

Selective breeding

Mankind has used selective breeding to produce many of the plants and animals that we rely on for food. Organisms that have the desired characteristics are allowed to breed, others are prevented from breeding. Over many generations, the population changes and shows the characteristics that are wanted. This also has the effect of limiting the variation of alleles in the population and therefore the genetic diversity.

The founder effect

The founder effect happens when a new population is formed by a small number of individuals. It might be a few birds landing on a newly formed island. The individuals will not have a great variety of alleles and so the population that they form will be restricted.

Genetic bottlenecks

A population may decrease in number due to natural events or man's influence. If the numbers become very small then the number of alleles will be limited. If the population recovers then the genetic diversity will be limited.

The numbers of Hawaiian goose dropped to about 30 birds in 1950 due to a number of factors such as over-hunting. Their numbers are starting to recover but their genetic diversity is really low.

Genetic variation in bacteria

Bacteria can reproduce every 20 minutes in ideal conditions.

Natural selection usually takes many thousands of years to produce changes, but in bacterial populations, changes can occur more rapidly. This is because they reproduce so quickly.

Mutations can occur to the DNA of bacteria when the DNA is copied prior to cell division. These mutations are passed on to the new bacteria and this is called **vertical gene transmission**. The new alleles may give the bacteria a selective advantage.

Although bacteria do not reproduce sexually, they can transfer DNA from one bacterium to another. This is called **conjugation**. It happens like this:
- One bacterial cell produces a thin tube from the cell membrane.
- This attaches to another bacterium.
- This allows small rings of DNA called plasmids to be transferred.

The passing on of genes by conjugation is called **horizontal gene transmission**.

Antibiotic resistance

One example of the importance of genetic variation in bacteria is their resistance to antibiotics.

> Antibiotics are substances produced by living organisms that destroy microorganisms or inhibit their growth.
>
> **KEY POINT**

Antibiotics prevent bacteria from making the cell wall properly. This means that when the bacteria take up water by osmosis the cell wall is too weak and bursts. The bacteria then die. This is called **osmotic lysis**.

Random mutations have occurred that allow bacteria to be resistant to the action of antibiotics like penicillin. These bacteria are more likely to survive and pass on the resistance by vertical gene transmission. This has produced strains that are resistant to certain antibiotics.

More worryingly, the genes can be passed from one strain to another by horizontal gene transmission. This has produced strains of bacteria such as MRSA and TB bacteria that are resistant to many different antibiotics.

Sample question and model answer

The quagga is an extinct mammal that looked like a zebra but had stripes only on its head, neck and forebody. The quagga lived in Africa until the last animal was shot in the late 1870s. The last specimen died in 1883 in a zoo.

The quagga, named *Equus quagga*, was originally thought to be a separate species to the plains Zebra.

Over the last fifty years or so, the markings of many individual zebras have been recorded by scientists.

Because of the great variation in stripe patterns, taxonomists were left with a problem. Was the quagga a separate species or was it a type of zebra? The quagga was the first extinct creature to have its DNA studied. Recent research using DNA from preserved specimens has demonstrated that the quagga was not a separate species at all, but a variety of the plains zebra, *Equus burchelli*.

Questions about natural selection always have similar mark schemes. Just make sure that you apply this mark scheme to the particular situation.

(a) What is the name of the genus that both the plains zebra and the quagga belong to? [1]

Equus

(b) What type of variation does the stripe pattern in zebras demonstrate? [1]

Continuous variation

(c) How could scientists use the DNA to investigate the relationship between the zebra and the quagga? [2]

Use DNA hybridisation.
The DNA is split into single strands and then scientists see how well it binds with zebra DNA strands.

(d) The plains zebra may have more stripes than the quagga because the stripes provide better camouflage on the plains.
How might this pattern have developed by natural selection? [3]

Zebras show variation and are born with different patterns of stripes.
More stripes make the zebra better camouflaged and so more likely to survive.
These zebras survive, breed and pass on the genes for more stripes.

Practice examination questions

1 The following key distinguishes between the five kingdoms.

1	Organisms without membrane bound organelles	A
	Organisms with membrane bound organelles	GO TO 2
2	Organisms have hyphae	B
	Organisms do not have hyphae	GO TO 3
3	Organisms unicellular or colonial	C
	Organisms not unicellular or colonial	GO TO 4
4	Organisms multicellular and have thylakoid membranes in some cells	D
	Organisms multicellular and have no thylakoid membranes in any cells	E

Name kingdoms A, B, C, D and E [5]

2 (a) The song-thrush (*Turdus ericetorum*) and mistle-thrush (*Turdus viscivorus*) are in the same family, *Turdidae*. Large sections of their DNA are common to both species. Complete the table to classify both organisms. [3]

	Mistle-thrush	Song-thrush
Kingdom		
Phylum	Chordata	Chordata
	Aves	
	Passeriformes	Passeriformes
Genus		
Species		

[1]　　　　　[1]　　　　　[1]

(b) How is it possible to find out if two female animals are from the same species? [1]

3 Students wished to investigate whether a pond in a field was being affected by a nearby cattle feeding station.

They sampled the water in the pond using a net and caught the following organisms:

(a) Use the student' results to produce an index of diversity for the sample. [2]

(b) The students sampled another pond in the same way.
This pond was further from the feeding station and their results produced an index of diversity of 6.90.
What conclusions can be drawn from the students' results? [2]

Chapter 8

Human health and disease

The following topics are covered in this chapter:

- *Health and lifestyle*
- *Disease*

- *Immunity*
- *Assisting immunity*

8.1 Health and lifestyle

After studying this section you should be able to:

- *define health*
- *understand how coronary heart disease can be affected by lifestyle*
- *describe the specific effects of smoking tobacco*

LEARNING SUMMARY

How can we achieve good health?

AQA A   1.1

Good health is not just an absence of disease or infirmity. It is the physical, mental and social well-being of a person. The development of a healthy person begins in the uterus. It is important that the mother supplies the fetus with suitable nutrients for development, e.g. amino acids for proteins essential for healthy growth. The mother's diet is important for both her and the fetus.

Following birth, the emotional and social development are equally as important as physical development.

If a person is healthy then they may expect the following:

- an absence of disease
- an absence of pain
- to be fit and have good muscle tone

- an absence of stress
- to get along with other people in society
- to have a long life expectancy.

Coronary heart disease and lifestyle

AQA   1.5

Atherosclerosis is a major health problem caused by eating saturated fats. This circulatory disease may develop as follows:

- yellow fatty streaks develop under the lining of the **endothelium** on the inside of an artery
- the streaks develop into a fatty lump called an **atheroma**
- the atheroma is made from **cholesterol** (taken up in the diet as well as being made in the liver)
- dense **fibrous tissue** develops as the atheroma grows
- the endothelial lining can **split**, allowing blood to contact the fibrous atheroma
- the damage may lead to a blood clot and an artery can be blocked.

collagen fibres

endothelium
a healthy artery

atheroma
fat and fibres

clot   atheroma so big that blockage taking place

Remember that the clotting of blood can occur for other reasons. There may be damage at other positions around the body. Blockage of this type is **thrombosis**.

Increasing constriction of an artery caused by **atherosclerosis** and **blood clots** reduces blood flow and increases blood pressure. If the artery wall is considerably weakened then a bulge in the side appears, just like a weakened inner tube on a cycle tyre. There is a danger of bursting and the structure is known as an **aneurysm**.

It is possible for a blood clot formed at an atheroma to break away from its original position. It may completely block a smaller vessel, this is known as an **embolism**.

Coronary heart disease

If the artery which supplies the heart (coronary artery) is partially blocked, then there is a reduction in oxygen and nutrient supply to the heart itself. This is called **coronary heart disease** (CHD). The first sign is often **angina**, the main symptom being sharp chest pains. If total blockage occurs then **myocardial infarction** (heart attack) takes place.

Lifestyle and CHD

Many aspects of lifestyle influence the condition of the cardiovascular system. CHD is a multi-factorial disease and the risk of developing CHD depends on a number of factors.

Statistically people have a greater chance of avoiding CHD if they:

- consume a low amount of saturated fat in their diet
- do not drink alcohol excessively
- consume a low amount of salt
- do not smoke
- are not stressed most of the time
- exercise regularly.

New drugs called statins are being prescribed in an effort to reduce CHD. They are thought to work by decreasing LDL levels in the blood.

Saturated fats (see lipid structure pages 19–20) are found in large quantities in animal tissues. Eating large quantities of saturated fats seems to increase the risk of CHD. It seems that saturated fat increases the levels of **low-density lipoproteins** (LDLs) in the blood. LDLs transport cholesterol in the blood and high levels of LDLs seem to increase the risk of atheroma formation. Other fats such as polyunsaturated fats seem to increase the levels of **high-density lipoproteins** (HDLs). This seems to give some protection against CHD.

Exercise has a **protective effect** on the **heart** and **circulation**. Activities such as jogging, walking, swimming and cycling can:

- reduce the resting heart rate
- increase the strength of contraction of the heart muscle
- increase the stroke volume of the heart (the volume of blood which is propelled during the contractions of the ventricles).

High salt levels have been shown to increase blood pressure, therefore increasing the risk of damage to atheromas.

Smoking also increases blood pressure and makes the blood more likely to clot.

Progress check

(a) Name a specific substance in food which can result in atherosclerosis.

(b) Describe and explain the structural changes which take place in a blood vessel as atherosclerosis develops.

(c) How can the damage caused by an atheroma result in a heart attack?

(c) a blood clot forms which can block the blood vessel completely. Prevention of oxygen supply to the heart results in myocardial infarction (heart attack).

(b) yellow fatty streaks develop under the cells lining the inside of a blood vessel, the streaks develop into a lump known as an atheroma, the atheroma is made of cholesterol, dense fibrous tissue develops and the lining of the vessel can split.

(a) saturated fat

What are the dangers of smoking tobacco?

AQA   1.4

Each person has another choice to make, to smoke or not to smoke. The government health warning on every cigarette packet informs of health dangers but many young people go ahead and ignore the information.

Effects of tobacco smoking

- **Nicotine** is the active component in tobacco which **addicts** people to the habit.
- **Tars** coat the alveoli which **slows down exchange** of carbon dioxide and oxygen. If less oxygen is absorbed then the smokers will be less active than their true potential.
- **Cilia** lining bronchial tubes are coated then **destroyed**, reducing the efficiency in getting rid of pollutants which enter the lungs. These pollutants include the cigarette chemicals themselves.
- **Carbon monoxide** from the cigarette gases **combines with haemoglobin** of red blood cells rather than oxygen. This **reduces oxygen transport** and the smokers become less active than their potential. Ultimately it may lead to heart disease.
- The **bronchi** and **bronchioles** become **inflamed**, a condition known as **bronchitis**. This causes irritating fluid in the lungs, **coughing** and increased risk of heart disease. A number of bronchitis sufferers die each year.
- The walls of the alveoli break down reducing the surface area for gaseous exchange. Less oxygen can be absorbed by the lungs, leaving the **emphysema** sufferers extremely breathless. They increase their breathing rate to compensate but still cannot take in enough oxygen for a healthy life. Chronic emphysema sufferers need an oxygen cylinder to prolong their life.
- **Blood vessel elasticity is reduced** so that serious damage may occur. Ultimately a heart attack can follow.
- The **carcinogens** (cancer-causing chemicals) of the tobacco can result in **lung cancer**. **Malignant growths** in the lungs develop uncontrollably and cancers may spread to other parts of the body. Death often follows. Smokers have a greater risk of developing other cancers than non-smokers, e.g. more smokers develop cervical cancer.

Even non-smokers can develop any of the above symptoms, but the probability of developing them is increased by smoking. Being in a smoky atmosphere each day also increases the chances.

> Learn the characteristics of each disease carefully. There are so many consequences of smoking that you may well mix them up.

The role of statistics

The government health warning on cigarette packets informs people of the risks of smoking. The WHO (World Health Organisation), governments and local authorities have collected statistics on many diseases over the years. These are used in education packs and posters to warn of risk factors. People can take precautions and use the information to avoid health dangers and take advantage of vaccination programmes.

Where education is not successful then related diseases follow. This acts as a drain on the National Health Service. Many operations which would have been unnecessary are performed to save people's lives, e.g. where coronary blood vessels are dangerously diseased a by-pass operation is the answer. The ideal situation is that education is successful, but realistically the aim is to balance prevention and cure.

8.2 Disease

After studying this section you should be able to:

- *define disease*
- *understand how infectious diseases are transmitted*

What is a disease?

AQA   1.1

A disease is a **disorder** of a tissue, organ or system of an organism. As a result of a disorder, **symptoms** are evident. Such symptoms could be the failure to produce a particular digestive enzyme, or a growth of cells in the wrong place. Normal bodily processes may be disrupted, e.g. efficient oxygen transport is impeded by the malarial parasite, *Plasmodium*.

Different types of disease

Infectious disease by pathogens
Pathogens attack an organism and can be passed from one person to another. Many pathogens are spread by a **vector** which carries them from one organism to another without being affected itself by the disease. Pathogens include bacteria, viruses, protozoa, fungi, parasites and worms.

> Most exam candidates recall that pathogens are responsible for disease. However, there are more causes of disease! If a question asks for different types of disease then giving a range of pathogens will not score many marks. Give genetic diseases, etc.

Genetic diseases
These can be passed from parent to offspring, e.g. haemophilia and cystic fibrosis.

Dietary related diseases
These are caused by the foods that we eat. Too much or too little food may cause disorders, e.g. obesity or anorexia nervosa. Lack of vitamin D causes the bone disease rickets, the symptoms of which are soft weak bones which bend under the body weight.

Environmentally related diseases
Some aspects of the environment disrupt bodily processes, e.g. as a result of nuclear radiation leakage, cancer may develop.

> An **auto-immune disease** may be **environmentally** caused, e.g. the form of leukaemia where phagocytes destroy a person's red blood cells may be caused by radiation leakage.

Auto-immune disease
The body in some way attacks its own cells so that processes fail to function effectively.

How are infectious diseases transmitted?

The pathogens which cause infectious diseases are transmitted in a range of ways.

- Direct contact – sexual intercourse enables the transmission of syphilis bacteria; a person's foot which touches a damp floor at the swimming baths can transfer the Athlete's foot fungus.

> Be prepared to answer questions about diseases not on your syllabus. The examiners will give data and other information which you will need to interpret. Use your knowledge of the principles of disease transmission, infection, symptoms and cure.

- Droplet infection – a sneeze propels tiny droplets of nasal mucus carrying viruses such as those causing influenza.
- Via a vector – if a person with typhoid bacteria in the gut handles food the bacteria can be passed to a susceptible person.
- Via food or water – chicken meat kept in warm conditions encourages the reproduction of *Salmonella* bacteria which are transferred to the human consumer, who has food poisoning as a result.
- Via blood transfusion – as a result of receiving infected blood a person can contract AIDS.

Some infectious diseases have serious consequences to human life. The incidence of infectious diseases may vary according to the climate of the country, the presence of vectors, the social behaviour of people and other factors.

Disease file – cholera

Cause of disease

Vibrio cholerae (bacteria) in the faeces or vomit of a human sufferer or human carrier which contaminate water supplies.

Transmission of microorganism

Contaminated water spreads the bacteria. Poor sanitary behaviour of people who are carriers and those who have contracted the disease are responsible. Faeces enters rivers which may be used for bathing, drinking, or irrigation. The bacteria survive outside the human body for around 24 hours. They can also contaminate vegetables and can be passed to a person in this way.

Outline of the course of the disease and symptoms

The bacteria reach the intestines where they breed. They secrete a toxin which increases secretion of chloride ions into the intestine, giving severe diarrhoea. Death is a regular consequence, due to dehydration, but some people do recover.

Prevention

Education about cleanliness and sewage treatment. Good sanitation is vital. Suitable treatment of water to be consumed by people, e.g. chlorination which kills the bacteria. Use of disinfectant also kills the bacteria. Early identification of an outbreak followed by control.

Cure

Tetracycline antibiotics kill organisms in the bowel. Immunisation is not very effective. It will help some individuals but not stop them from being carriers, so epidemics are still likely.

Oral rehydration therapy is usually used to combat the dehydration caused by diarrhoea and vomiting.

Simply drinking water is ineffective because the electrolyte losses also need replacing. The standard treatment is to restore fluids intravenously with water and salts. This requires trained personnel and materials which are not always available.

However, it was discovered that the body can absorb a simple solution containing both **sugar and salt**. The dry ingredients can be mixed and packaged, and then the solution can be given by people with minimal training. In cholera, the toxin interferes with one transport system in the intestine lining but the sodium-glucose co-transporter still operates. This allows absorption of glucose, sodium and water.

Oral rehydration can therefore be accomplished by drinking frequent small amounts of an **oral rehydration solutions**.

Disease file – tuberculosis

AQA 1.4

Cause of disease

Mycobacterium tuberculosis (bacterium) via droplet infection.

Transmission of microorganism

Coughs and sneezes of sufferers spread tiny droplets of moisture containing the pathogenic bacteria. People then inhale these droplets and may contract the disease.

Outline of the course of the disease and symptoms

The initial attack takes place in the lungs. The alveoli surfaces and capillaries are vulnerable and lesions occur. Some epithelial tissues begin to grow in number but these cannot carry out gaseous exchange. Inflammation occurs which stimulates painful coughing. Intense coughing takes place which can cause bleeding. There is much weight loss. Weak groups of people, like the elderly, or those underweight are more prone to the disease.

Prevention

Mycobacterium bovis causes tuberculosis in cattle. It can be passed to humans via milk. It causes an intestinal complaint in humans. It is important that cows are kept free of *M. bovis* by antibiotics.

The BCG vaccination is the injection of a weakened form of this microbe. This vaccination stimulates antibodies which are effective against both *M. tuberculosis* and *M. bovis*.

Mass screening using **X-rays** can identify 'shadows' in those people with scar tissue in the lungs.

Sputum testing identifies the presence of the bacteria in sufferers. Sufferers can be treated with antibiotics. Once cured they cannot pass on the pathogen so an epidemic may be prevented.

Skin testing is used. Antigens from dead *Mycobacteria* are injected just beneath the skin. If a person has been previously exposed to the organism then the skin swells which shows that they already have resistance, i.e. they have antibodies already. Anyone whose skin does not swell up is given the **BCG vaccination**. This contains attenuated *Mycobacterium bovis* to stimulate the production of antibodies against both *M. bovis* and *M. tuberculosis*.

Cure

Use of antibiotics such as streptomycin.

8.3 Immunity

After studying this section you should be able to:

• *describe and explain the action of the body's immune system*

LEARNING SUMMARY

The roles of the white blood cells (leucocytes)

AQA 1.6

There are a number of different types of **leucocytes**. Most are produced from **stem cells** in the **bone marrow**. Different stem cells follow alternative maturation procedures to produce a range of leucocytes. Leucocytes have the ability to recognise self chemicals and non-self chemicals. Only where non-self chemicals are recognised will a leucocyte respond. Proteins and polysaccharides are typical of the complex molecules which can trigger an immune response.

Phagocytes

Phagocytes can move to a site of infection through capillaries, tissue fluid and lymph as well as being found in the plasma. They move towards pathogens which they destroy by the process of phagocytosis. This is often called engulfment and involves the surrounding of a pathogen by pseudopodia to form a food vacuole. Hydrolytic enzymes from lysosomes complete the destruction of the pathogen.

Neutrophils are one type of phagocyte. Proteins in plasma called **opsonins** attach to a pathogen. These opsonins enable the phagocyte to engulf the pathogen.

Macrophages are another type of phagocyte which work alongside T-lymphocytes.

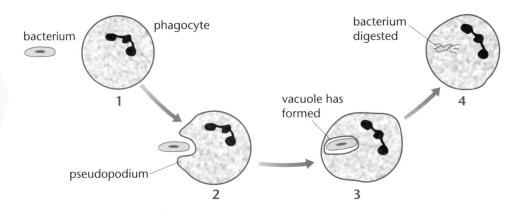

What is an antigen?

AQA A 1.6

As an individual grows and develops, complex substances such as proteins and polysaccharides are used to form cellular structures. Leucocytes identify these substances in the body as 'self' substances. They are ignored as the leucocytes encounter them daily. 'Non-self substances', e.g. foreign proteins which enter the body, are immediately identified as 'non-self'. These are known as **antigens** and trigger an immune response.

Lymphocytes

White blood cells (leucocytes) constantly check out proteins around the body. Foreign protein is identified and attack is stimulated.

In addition to phagocytes, there are other leucocytes called lymphocytes. There are two types of lymphocyte, **B-lymphocytes** and **T-lymphocytes**

B-lymphocytes begin development and mature in the bone marrow. They produce antibodies, known as the **humoral response**

T-lymphocytes work alongside phagocytes known as macrophages; this is known as the **cell-mediated response**. A macrophage engulfs an antigen. This antigen remains on the surface of the macrophage. T-lymphocytes respond to the antigen, dividing by mitosis to form a range of different types of T-lymphocyte cells.

- **Killer T-lymphocytes** adhere to the pathogen, secrete a toxin and destroy it.
- **Helper T-lymphocytes** stimulate the production of antibodies.
- **Suppressor T-lymphocytes** are inhibitors of the T-lymphocytes and plasma cells. Just weeks after the initial infection, they shut down the immune response when it is no longer needed.
- **Memory T-lymphocytes** respond to an antigen previously experienced. They are able to destroy the same pathogen before symptoms appear.

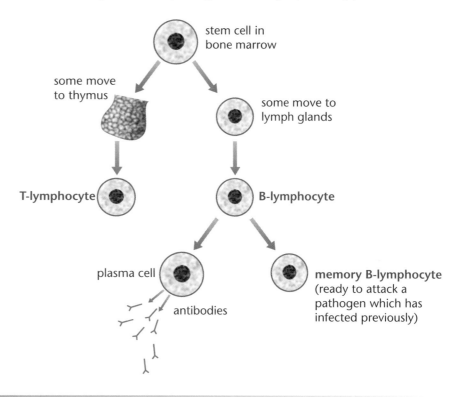

How do antibodies destroy pathogens?

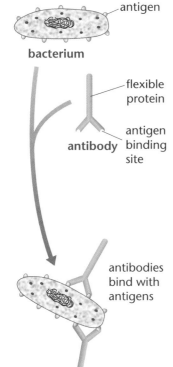

The diagram on the left shows antibodies binding to antigens. The descriptions below show what can happen immediately after the binding takes place.

There are three main ways in which antibodies destroy pathogens.

- **Precipitation**, by linking many antigens together. This enables the phagocytes to engulf them.
- **Lysis**, where the cell membrane breaks open, killing the cell.
- **Neutralisation** of a chemical released by the pathogen, so that the chemical is no longer toxic.

8.4 Assisting immunity

After studying this section you should be able to:

- *describe the origin of different types of immunity*
- *explain how immunity can be stimulated by vaccinations*

Stimulating the immune system

AQA ▷ 1.6

Newborn babies are naturally protected against many diseases such as measles and poliomyelitis. This is because they have received antibodies from their mother in two ways:

- across the placenta
- via breast feeding.

> This is one major advantage of breast feeding over bottle feeding of babies.

This type of immunity is called **passive natural immunity**. It is possible to give people injections of antibodies that have been made by another person or animal. This type of immunity is called **passive artificial** immunity.

> Passive immunity does not last very long. This is because the antibodies do not persist in the blood for very long.

K E Y
P O I N T

Active immunity

Once a person has recovered from certain diseases, e.g. measles, they rarely contract that disease again. This is because the memory cells formed from B-lymphocytes can survive in the blood for many years. If the antigen reappears, they can rapidly produce a clone of antibody producing cells. This **secondary response** is rapid and larger than the **primary response** and the antigen is rapidly destroyed.

This type of immunity is called **active natural immunity**.

Vaccines consist of dead or weakened (**attenuated**) forms of pathogens. They stimulate the production of memory cells so that the person develops active artificial **immunity**.

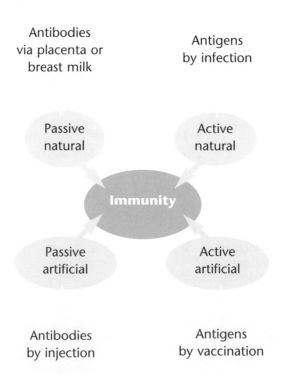

Antibodies
via placenta or
breast milk

Antigens
by infection

Passive
natural

Active
natural

Immunity

Passive
artificial

Active
artificial

Antibodies
by injection

Antigens
by vaccination

Unfortunately some pathogens show **antigenic variation**. This change in antigens renders vaccinations ineffective after a while and allows diseases such as influenza to strike many times in slightly different forms.

Scientists are now using antibody producing cells that have been fused with tumour cells. This produces cells that make large quantities of one type of antibody. They are called monoclonal antibodies. Because of their ability to target particular antigens, they might be useful in delivering drugs straight to target cells.

Sample question and model answers

The graph below shows the relative numbers of antibodies in a person's blood after the vaccination of attenuated viruses. Vaccinations were given on day 1 then 200 days later.

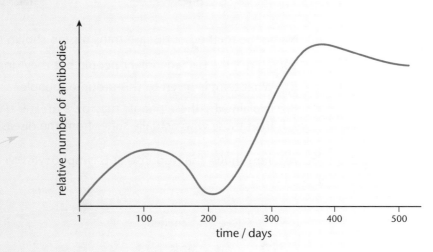

In examinations you are regularly given graphs. Make sure that you can link the idea being tested. This should help you recall all of the important concepts needed. All you need to do after this is **apply** your knowledge to the given data.

(a) Why is it important that viruses used in vaccinations are attenuated? [1]

If they were active then the person would contract the disease.

(b) Suggest **two** advantages of giving the second vaccination. [2]

A greater number of antibodies were produced.
The antibodies remain for much longer after the second vaccination.

(c) Which cells produced the antibodies during the primary response? [1]

B–lymphocytes.

(d) Why was there no delay in the secondary response to vaccination? [3]

Because the first vaccination had already been given, memory B–lymphocytes had been produced which respond to the viruses more quickly.

(e) Describe how a virus stimulates the production of antibodies? [3]

Antigen in the protein 'coat' or capsomere stimulate the B–lymphocytes.

If you gave B-lymphocytes as a response it would be wrong! B-lymphocytes secrete antibodies.

(f) Apart from producing antibodies, outline TWO different ways that the body uses to destroy microorganisms. [2]

Phagocytes by engulfment; T–lymphocytes attach to microorganisms and destroy them.

Practice examination questions

1 The illustration shows how the bacterium which causes TB can be transmitted from one person to another.

(a) Name the method of disease transmission shown in the illustration.   [1]

(b) Sometimes the bacteria infect people but they do not develop symptoms.

  (i) What term is given to this group of people?   [1]

  (ii) Explain why these people may be a greater danger to a community than those who actually suffer from the disease.   [2]

2 The diagram shows the response of B-lymphocytes to a specific antigen.

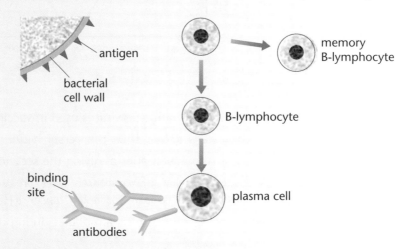

(a) (i) A plasma cell is bigger than a B-lymphocyte.
     Suggest an advantage of this.   [1]

  (ii) Describe the precise role of antibodies in the immune response.   [3]

  (iii) What is the advantage of memory B-lymphocytes?   [2]

(b) What is an auto-immune disease? Give an example.   [2]

(c) A person contracts the virus which causes the common cold.
    Suggest why their lymphocytes may fail to destroy the pathogen.   [1]

Practice examination answers

Chapter 1 Biological molecules

1 (a) RCOOH   HOCH$_2$
       RCOOH + HOCH
       RCOOH   HOCH$_2$
       fatty acids glycerol   [2]

   (b) Emulsion test: add the sample to ethanol and mix;
decant or pour into water;
if a fat is present a white emulsion forms on the
surface.   [3]
                      [Total: 5]

2 (a) peptide bond/peptide link   [1]
   (b) –COOH/carboxylic acid   [1]
   (c) primary structure; amino acids in a chain [2]
                      [Total: 4]

Chapter 2 Cells

1 (a) A = phospholipid   B = protein   [2]
   (b) Acts as a channel/pore; to allow molecules in or
out of the cell.   [1]
   (c) One end is hydrophobic and the other is
hydrophilic; Hydrophilic tails 'hide' in the centre of
the membrane;   [2]
                      [Total: 5]

2 (a) C = Golgi body
      B = Centrioles
      A = Cell membrane
      E = Mitochondria
      D = Rough endoplasmic reticulum   [5]
   (b) Correct measure of width nucleus (~2.0 cm)
convert to micrometres (20 000 µm)
divide size by 5 000 (4 µm)   [3]
   (c) Liver cells carry out many functions;
Need large amounts of energy/ATP   [2]
                      [Total: 10]

3 (a) cell wall (not cellulose); no true nucleus; no
mitochondria; plasmids [any two for 2 marks] [2]
   (b) mitochondria; nucleus; Golgi body; large
ribosomes/rough endoplasmic reticulum
[any two for 2 marks]   [2]
                      [Total: 4]

4 (a) lens X = projector; lens Y = objective   [2]
   (b) electrons would collide with air molecules [1]
   (c) (i) artefact   [1]
        (ii) ignore the artefact, because it is not part of
            normal structure and is only present
            due to preparation of the specimen. [1]
                      [Total: 5]

Chapter 3   Enzymes

1 (a) lock and key – the substrate is a similar shape to the active site; it fits in and binds with the active site like a key (substrate) fitting into a lock (active site); induced fit – the substrate is not a matching 'fit' for the active site, but as the substrate approaches, the active site changes into an appropriate shape.

(b) reversible – enters active site but will come out again irreversible – binds permanently with enzymes   [4]
[Total: 4]

2 • The tertiary structure of the enzyme is responsible for the further folding of the protein;
• this gives the shape of the active site;
• the active site in amylase is specific to starch, lipid is unable to bind to the active site of amylase, therefore amylase cannot break down lipid.   [3]
[Total: 3]

3 (a) starch   [1]
(b) thermostable enzymes are effective at high temperatures   [1]
[Total: 2]

4 (a) The substrate molecule collides with the active site of the enzyme; as it approaches, the active site changes shape to become compatible with the substrate shape.   [2]

(b) Non-competitive inhibitor molecule binds with part of enzyme other than the active site; as a result the active site changes shape; so the substrate can no longer bind with the active site.   [3]
[Total: 5]

5 (a) NH_2 (amino group)   [1]

(b) The enzymes then have more ends to attack.   [2]
[Total: 3]

Chapter 4   Exchange

1 (a) (i) No change in size because the water potential inside the cell equals the water potential of the solution outside the cell.   [1]
(ii) The water potential of the solution outside the cell is more negative than the water potential inside the cell.   [1]
(iii) The water potential of the solution outside the cell is less negative than the water potential inside the cell.   [1]

(b) osmosis   [1]
[Total: 4]

2 (a) They both use a protein carrier molecule.   [1]

(b) Active transport needs energy or mitochondria, whereas facilitated diffusion does not.
OR Active transport allows molecules to move from a lower concentrated solution to a higher concentrated solution.
OR Active transport allows molecules to move against a concentration gradient.   [1]
[Total: 2]

3 • a flat, thin blade allows maximum light absorption;
• the leaf has a waxy cuticle to reflect excess light but allow entry of enough light for photosynthesis;
• the mesophyll cells have many chloroplasts to absorb the maximum amount of light;
• the palisade cells pack closely together to absorb the maximum amount of light;

• guard cells open stomata to allow carbon dioxide in and oxygen out during photosynthesis;
• air spaces in the mesophyll store lots of carbon dioxide for photosynthesis;
• xylem of the vascular bundles bring water to the leaf for photosynthesis;
• phloem takes the carbohydrate away from the leaf after photosynthesis.
[any 6 points]   [6]
[Total: 6]

4 (a) alveoli have a very high surface area; they are very close to many capillaries; capillaries are one cell thick/very thin/have squamous epithelia; they are kept damp which facilitates diffusion.   [4]

(b) • there are many gill filaments, which give a very high surface area;
• the gill filaments are very thin;
• they have many capillaries;
• they have gill plates which increase the surface area further;
• the blood and water directions are opposite which maximises diffusion/they use countercurrent flow to maximise diffusion.
[any 4 points]   [4]
[Total: 8]

Chapter 5 Transport

1 (a) to the body core [1]

   (b) less blood reaches the superficial capillaries of the skin; so less heat is lost by conduction, convection and radiation; blood in the body core better insulated by the adipose layer of the skin [4]
   [Total: 5]

2 (a) 60 x 120 x 5 = 36 000 ml / 36 litres [2]

   (b) • blood is transported more quickly;
        • more oxygen taken up at the lungs/more carbon dioxide excreted at the lungs;
        • more oxygen reaches the muscles;
        • more glucose reaches the muscles;
        • so muscles contract more effectively.
        [any 4 points] [4]

   (c) (i) • slower breathing rate;
            • more alveoli accessed for exchange;
            • intercostal muscles more effective.
            [any 2 points] [2]
        (ii) • improved muscle tone;
            • greater muscular strength;
            • more capillaries in muscles. [any 2 points] [2]
            [Total: 10]

3 (a) The amount of water lost by transpiration is exactly matched by the amount taken up by the leaf. [1]

   (b) (i) xylem [1]
        (ii) fill the potometer under water; operate the valve to get rid of air bubbles; when removing the leaf from the tree, the stalk or petiole must be put in water immediately. [2]

   (c) Volume of water = πr^2 x 32 mm x 60
        $= \dfrac{22}{7}$ x 1 x 1 x 32 x 60
        = 6034.3 mm³ [3]
        [Total: 7]

4 • extensive root system to absorb maximum water;
   • large amount of water storage in leaves;
   • thick cuticle;
   • low numbers of stomata/sunken stomata;
   • hairs to reduce turbulence. [any 3 points] [3]
   [Total: 3]

5 (a) SAN/Sinoatrial node [1]

   (b) (i) slows heart rate [1]
        (ii) speeds up the rate [1]
        (iii) speeds up the rate [1]
        [Total: 4]

Chapter 6 Genes and cell division

1 (a) metaphase [1]
   (b) 4 [2]
        [Total: 3]

2 (a) adenine and thymine are similar proportions because adenine binds with thymine; cytosine and guanine are similar proportions because cytosine binds with guanine [2]

   (b) They should be identical in number but the scientists were operating at the limits of instrumentation. [1]

   (c) Organic bases form the codes for different amino acids. Different sequences of amino acids form the different proteins specific to a species. [2]
        [Total: 5]

Chapter 7 Classification and biodiversity

1 A = Prokaryotae
B = Fungi
C = Protoctista
D = Plantae
E = Animalia [5]

[Total: 5]

2 (a)

	Mistle-thrush	*Song-thrush*
Kingdom	**Animalia**	**Animalia**
Phylum	Chordata	Chordata
Class	Aves	**Aves**
Order	Passeriformes	Passeriformes
Family	**Turdidae**	**Turdidae**
Genus	**Turdus**	**Turdus**
Species	**viscivorus**	**ericetorum**

[3]

(b) disruptive selection [1]

[Total: 4]

3 (a) N = 20

$$d = \frac{20 \times 19}{(3 \times 2) + (6 \times 5) + (2 \times 1) + (1 \times 0) + (4 \times 3) + (3 \times 2) + (1 \times 0)}$$

$$= \frac{380}{56} = 6.79$$ [2]

(b) Biodiversity was lower in pond near feeding station. Difference may not be significant. [2]

[Total: 4]

Chapter 8 Human health and disease

1 (a) droplet infection [1]

(b) (i) carriers [1]
(ii) we do not know that they carry the pathogen as they display no symptoms, so the people do not avoid contact and pass on the bacteria [2]

[Total: 4]

2 (a) (i) ability to secrete more antibodies [1]
(ii) the antibodies have specific receptor sites which bind with the antigens; they have a flexible protein which changes angle to fit the antigens; antibodies result in the destruction of the antigen in some way/ neutralise toxin/cluster around antigens then cause precipitation/cause agglutination [3]
(iii) are produced when body first exposed to antigen; remain in body to react quickly when exposed to same antigen again [2]

(b) when the immune system attacks the person's own cells; pernicious anaemia/rheumatoid arthritis [2]

(c) the influenza virus often mutates so lymphocytes take longer to produce antibodies [1]

[Total: 9]

Index

Index

Revise A2

AQA Biology

John Parker & Ian Honeysett

Contents

Contents

Specification list

The specification labels on each page refer directly to the units in the exam board specification, i.e. AQA 4.1 refers to Unit 4, section 1.

AQA Biology

UNIT	SPECIFICATION TOPIC	CHAPTER REFERENCE	STUDIED IN CLASS	REVISED	PRACTICE QUESTIONS
Unit 4	Populations	7.1, 7.2			
	ATP	1.1			
	Photosynthesis	1.2			
	Respiration	1.3			
	Energy transfer	8.1, 8.2			
	Nutrient cycles	8.3, 8.4			
	Succession	7.1, 7.2, 7.3			
	Inheritance	4.1, 4.2			
	Selection and speciation	5.2			
Unit 5	Stimuli	2.1, 2.3, 2.4, 2.5			
	Coordination	2.2, 2.5, 3.1			
	Muscles	2.4			
	Homeostasis	3.1, 3.2, 3.3			
	Protein synthesis	3.1, 3.4, 4.1, 5.1			
	Gene expression	4.1, 5.1			
	Gene technology	6.1			

Examination analysis

Unit 4
Examination paper (100 UMS) (75 raw marks)
6-9 short answer questions plus 2 longer questions (a short comprehension and a short structured essay)
$1\frac{1}{2}$ hours
16.67% of the total A level marks

Unit 5
Examination paper (140 UMS) (100 raw marks)
8–10 short answer questions plus 2 longer questions (a data-handling question and a synoptic essay – choice of 1 out of 2)
$2\frac{1}{4}$ hours
23.33% of the total A level marks

Unit 6
A2 Centre Assessed Unit (60 UMS) (50 raw marks)
Practical Skills Assessment (PSA) 6 marks
Investigative Skills Assignment (ISA) 44 marks
10% of the total A level marks

The AS/A2 Level Biology course

All Biology GCE A level courses currently studied are in two parts: AS and A2, with three separate units in each.

Some of the units are assessed by written papers, externally marked by the Awarding Body. Some units involve internal assessment of practical skills (subject to moderation).

Each Awarding Body has a common core of subject content in AS and A2. Beyond the common core material, the Awarding Bodies have included more varied content. This study guide contains the common core material and the additional material that is relevant to the AQA A2 specification.

In using this study guide, some students may have already completed the AS part of the course. Knowledge of AS is assumed in the A2 part of the course. It is therefore important to revisit the AS information when preparing for the A2 examinations.

What are the differences between AS and A2?

There are three main differences:

(i) A2 includes the more **demanding** concepts. (Understanding will be easier if you have completed the AS Biology course as a 'stepping stone'.)

(ii) There is a much greater emphasis on the skills of **application** and **analysis** than in AS. (Using knowledge and understanding acquired from AS is essential.)

(iii) A2 includes a substantial amount of **synoptic** material. (This is the drawing together of knowledge and skills across the modules of AS and A2. Synoptic investigative tasks and questions involving concepts across the specification are included.)

How will you be tested?

Assessment units

A2 Biology comprises three units. The first two units are assessed by examinations.

The third component involves either centre assessed practical work or AQA assessed practical work. This tests practical skills. If it is marked by your teacher, the marks can be adjusted by moderators appointed by AQA.

Tests are taken at two specific times of the year, January/February and June. If you are disappointed with a unit result, you can resit each unit any number of times. It can be an advantage to you to take a unit test at the earlier optional time because you can re-sit the test. The best mark from each unit will be credited and the lower marks ignored.

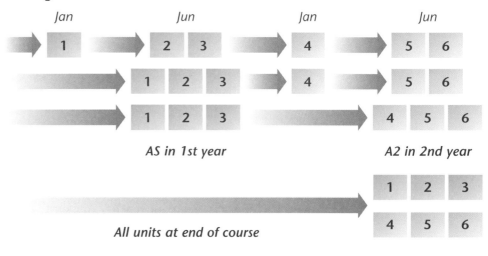

AS in 1st year

A2 in 2nd year

All units at end of course

A2 and synoptic assessment

Most students who study A2 have already studied to AS Level. There are three further units to be studied.

Every A Level specification includes synoptic assessment at the end of A2. Synoptic questions draw on the ideas and concepts of earlier units, bringing them together in holistic contexts. Examiners will test your ability to inter-relate topics through the complete course from AS to A2. (See the synoptic chapter page 101).

What skills will I need?

For A2 Biology, you will be tested by assessment objectives: these are the skills and abilities that you should have acquired by studying the course. The assessment objectives are shown below.

Knowledge with understanding

- recall of facts, terminology and relationships
- understanding of principles and concepts
- drawing on existing knowledge to show understanding of the responsible use of biological applications in society
- selecting, organising and presenting information clearly and logically

Application of knowledge and understanding, and evaluation

- explaining and interpreting principles and concepts
- interpreting and translating, from one to another, data presented as continuous prose or in tables, diagrams and graphs
- carrying out relevant calculations
- applying knowledge and understanding to familiar and unfamiliar situations
- assessing the validity of biological information, experiments, inferences and statements

You must also present arguments and ideas clearly and logically, using specialist vocabulary where appropriate. Remember to balance your argument!

Investigative and practical skills

One of the three A2 units (and one of the AS units) is designed to test your investigative and practical skills. Each unit is made up of two parts:
- Practical Skills Assessment (PSA)
- Investigative Skills Assessment (ISA)

The PSA involves your teacher marking you on a scale of 0 to 6 based on your practical work throughout the course. They will be looking to see if you can…
- Follow instructions
- Select and use equipment
- Carry out practical work in an organised and safe way.

The ISA carries 44 marks and is made up of three tasks:
- First you collect some practical data.
- Then you process the data.
- This is followed by a written paper relating to the experiment.

These ISA tasks are provided by AQA and you will carry them out under controlled conditions, being supervised by your teacher. Your teacher then marks them using a mark scheme provided by AQA. A task cannot be repeated but you may be given the opportunity to do more than one type of each task. Your best mark on each will count. (There is an alternative to the ISA, which is provided and marked by AQA. Only one of these is provided each year.)

Different types of questions in A2 examinations

Questions in AS and A2 Biology are designed to assess a number of assessment objectives. For the written papers in Biology the main objectives being assessed are:

- recall of facts, terminology and inter-relationships
- understanding of principles and concepts and their social and technological applications and implications
- explanation and interpretation of principles and concepts
- interpreting information given as diagrams, photomicrographs, electron micrographs tables, data, graphs and passages
- application of knowledge and understanding to familiar and unfamiliar situations.

In order to assess these abilities and skills a number of different types of question are used.

In A2 Level Biology unit tests these include short-answer questions and structured questions requiring both short answers and more extended answers, together with free-response and open-ended questions.

Short-answer questions

A short-answer question will normally begin with a brief amount of stimulus material. This may be in the form of a diagram, data or graph. A short-answer question may begin by testing recall. Usually this is followed up by questions which test understanding. Often you will be required to analyse data. Short-answer questions normally have a space for your responses on the printed paper. The number of lines is a guide as to the number of words you will need to answer the question. The number of marks indicated on the right side of the papers shows the number of marks you can score for each question part. Here are some examples. (The answers are shown in blue).

The diagram below shows a gastric pit.

(a) (i)   Label cell X   (1)
          oxyntic cell

    (ii) What is secreted by cell X?   (1)
          hydrochloric acid

(b) (i)   Protein enters the stomach. What must take place before the
          hydrolysis of the protein begins?   (2)
          Hydrochloric acid acts on pepsinogen, to produce pepsin

    (ii) After the protein has been hydrolysed, what is produced?   (1)
          polypeptides

Structured questions

Structured questions are in several parts. The parts are usually about a common context and they often progress in difficulty as you work through each of the parts. They may start with simple recall, then test understanding of a familiar or unfamiliar situation. If the context seems unfamiliar the material will still be centred around concepts and skills from the Biology specification. (If a student can answer questions about unfamiliar situations then they display understanding rather than simple recall.)

The most difficult part of a structured question is usually at the end. Ascending in difficulty, a question allows a candidate to build in confidence. Right at the end technological and social applications of biological principles give a more demanding challenge. Most of the questions in this book are structured questions. This is the main type of question used in the assessment of both AS and A2 Biology.

The questions set at A2 Level are generally more difficult than those experienced at AS Level. A2 includes a number of higher-level concepts, so can be expected to be more difficult. The key advice given by this author is:

- Give your answers in greater detail:

    Example: Why does blood glucose rise after a period without food?

    Answer: The hormone glucagon is produced ✗ (this is not enough for credit!)

    The hormone glucagon is produced which results in glycogen breakdown to glucose. ✓

- Look out for questions with a 'sting in the tail'. A2 structured questions are less straightforward, so look for a 'twist'.

When answering structured questions, do not feel that you have to complete one question before starting the next. Answering a part that you are sure of will build your confidence. If you run out of ideas go on to the next question. This will be more profitable than staying with a very difficult question which slows down progress. You can return at the end when you have more time.

Extended answers

In A2 and AS Biology, questions requiring more extended answers will usually form part of structured questions. They will normally appear at the end of a structured question and will typically have a value of 4 to 10 marks. Longer questions are allocated more lines, so you can use this as a guide as to how many points you need to make in your response. Often for an answer worth 10 marks the mark scheme would have around 12 to 14 creditable answers. You are awarded up to the maximum, 10 marks, in this instance.

On the Unit 5 paper, there is a synoptic essay question. This is worth 25 marks. This includes marks for the **quality of communication**.

Candidates are assessed on their ability to use a suitable style of writing, and organise relevant material, both logically and clearly. The use of specialist biological terms in context is also assessed. Spelling, punctuation and grammar are also taken into consideration.

Question

Explain how an action potential is transmitted across a synapse.

(Total 13 marks)

The action potential depolarises the synaptic knob. ✓ This causes the calcium channels to open. ✓ There is an inflow of Ca++ into the synaptic bouton. ✓ This causes the synaptic vesicles to move towards the pre-synaptic membrane. ✓ The vesicles then fuse with the pre-synaptic membrane. ✓ They then release the neurotransmitter into the synaptic cleft, ✓ by exocytosis. ✓ The neurotransmitter diffuses across the synaptic cleft and attaches to receptors on the post synaptic membrane. ✓ This results in a change in the permeability of post-synaptic membrane/Na+ gates opened. ✓ Sodium ions enter the post synaptic cell. ✓ This will depolarise the membrane ✓ and an action potential is initiated. ✓ Cholinesterase is released into the cleft to remove the transmitter. ✓

Communication mark ✓

Remember that mark schemes for extended questions often exceed the question total, but you can only be awarded credit up to the maximum. In response to this question the candidate would be awarded the maximum of 13 marks which included one communication mark. The candidate gave one more creditable response which was on the mark scheme, but had already scored a maximum. Try to give more detail in your answers to longer questions. This is the key to A2 success.

Stretch and Challenge

Stretch and Challenge is a concept that is applied to the structured questions in Unit 4 and 5 of the exam papers in A2. In principle, it means that sub-questions become progressively harder so as to challenge more able students and help differentiate between A and A* students.

Stretch and Challenge questions are designed to test a variety of different skills and your understanding of the material. They are likely to test your ability to make appropriate connections between different areas and apply your knowledge in unfamiliar contexts (as opposed to basic recall).

Exam technique

A2 builds from the skills and concepts acquired during the AS course. It will help you cope as the A2 concepts ascend in difficulty. The chapters explain the ideas in small steps so that understanding takes place gradually. The final aim, of complete understanding of major topics, is more likely.

Can I use the AS Biology Study Guide for A2?

YES! This will be particularly useful in answering synoptic questions that require direct knowledge of the AS topics.

What are examiners looking for?

Whatever type of question you are answering, it is important to respond in a suitable way. Examiners use instructions to help you to decide the length and depth of your answer. The most common words used are given below, together with a brief description of what each word is asking for.

Define

This requires a formal statement. Some definitions are easy to recall.

> Define the term transport.

> *This is the movement of molecules from where they are in lower concentration to where they are in higher concentration. The process requires energy.*

Other definitions are more complex. Where you have problems it is helpful to give an example.

> Define the term endemic.

> *This means that a disease is found regularly in a group of people, district or country.*

Use of an example clarifies the meaning. Indicating that malaria is invariably found everywhere in a country confirms understanding.

Explain

This requires a reason. The amount of detail needed is shown by the number of marks allocated.

> Explain the difference between resolution and magnification.

> *Resolution is the ability to be able to distinguish between two points whereas magnification is the number of times an image is bigger than an object itself.*

State

This requires a brief answer without any reason.

> State one role of blood plasma in a mammal.

> *Transport of hormones to their target organs.*

List

This requires a sequence of points with no explanation.

List the abiotic factors which can affect the rate of photosynthesis in pondweed.

carbon dioxide concentration; amount of light; temperature; pH of water

Describe

This requires a piece of prose which gives key points. Diagrams should be used where possible.

Describe the nervous control of heart rate.

The medulla oblongata ✓ of the brain connects to the sino–atrial node in the right atrium, wall ✓ via the vagus nerve and the sympathetic nerve ✓ the sympathetic nerve speeds up the rate ✓ the vagus nerve slows it down.✓

Discuss

This requires points both for and against, together with a criticism of each point. (**Compare** is a similar command word).

Discuss the advantages and disadvantages of using systemic insecticides in agriculture.

Advantages are that the insecticides kill the pests which reduce yield ✓ they enter the sap of the plants so insects which consume sap die ✓ the insecticide lasts longer than a contact insecticide, 2 weeks is not uncommon ✓

Disadvantages are that insecticide may remain in the product and harm a consumer e.g. humans ✓ it may destroy organisms other than the target ✓ no insecticide is 100% effective and develops resistant pests. ✓

Suggest

This means that there is no single correct answer. Often you are given an unfamiliar situation to analyse. The examiners hope for logical deductions from the data given and that, usually, you apply your knowledge of biological concepts and principles.

The graph shows that the population of lynx decreased in 1980. Suggest reasons for this.

Weather conditions prevented plant growth ✓ so the snowshoe hares could not get enough food and their population remained low ✓ so the lynx did not have enough hares (prey) to predate upon. ✓ The lynx could have had a disease which reduced numbers. ✓

Calculate

This requires that you work out a numerical answer. Remember to give the units and to show your working, marks are usually available for a partially correct answer. If you work everything out in stages write down the sequence. Otherwise if you merely give the answer and it is wrong, then the working marks are not available to you.

Calculate the Rf value of spot X. (X is 25 mm from start and solvent front is 100 mm)

$$Rf = \frac{distance\ moved\ by\ spot}{distance\ moved\ by\ the\ solvent\ front}$$

$$= \frac{25\ mm}{100\ mm} = 0.25$$

Outline

This requires that you give only the main points. The marks allocated will guide you on the number of points which you need to make.

Outline the use of restriction endonuclease in genetic engineering.

The enzyme is used to cut the DNA of the donor cell. ✓

It cuts the DNA up like this
$$ATGCCGAT = AT \quad + GCCGAT \checkmark$$
$$TACGGCTA \quad TACGGC \quad\quad TA$$

The DNA in a bacterial plasmid is cut with the same restriction endonuclease. ✓

The donor DNA will fit onto the sticky ends of the broken plasmid. ✓

If a question does not seem to make sense, you may have misread it. Read it again!

Some dos and don'ts

Dos

Do *answer the question*

No credit can be given for good Biology that is irrelevant to the question.

Do *use the mark allocation to guide how much you write*

Two marks are awarded for two valid points – writing more will rarely gain more credit and could mean wasted time or even contradicting earlier valid points.

Do *use diagrams, equations and tables in your responses*

Even in 'essay-style' questions, these offer an excellent way of communicating Biology.

Do *write legibly*

An examiner cannot give marks if the answer cannot be read.

Do *write using correct spelling and grammar. Structure longer essays carefully*

Marks are now awarded for the quality of your language in exams.

Don'ts

Don't *fill up any blank space on a paper*

In structured questions, the number of dotted lines should guide the length of your answer.

If you write too much, you waste time and may not finish the exam paper. You also risk contradicting yourself.

Don't *write out the question again*

This wastes time. The marks are for the answer!

Don't *contradict yourself*

The examiner cannot be expected to choose which answer is intended. You could lose a hard-earned mark.

Don't *spend too much time on a part that you find difficult*

You may not have enough time to complete the exam. You can always return to a difficult calculation if you have time at the end of the exam.

What grade do you want?

Everyone would like to improve their grades but you will only manage this with a lot of hard work and determination. You should have a fair idea of your natural ability and likely grade in Biology and the hints below offer advice on improving that grade.

For a Grade A

You will need to be a very good all-rounder.

- You must go into every exam knowing the work extremely well.
- You must be able to apply your knowledge to new, unfamiliar situations.
- You need to have practised many, many exam questions so that you are ready for the type of question that will appear.

The exams test all areas of the syllabus and any weaknesses in your Biology will be found out. There must be no holes in your knowledge and understanding. For a Grade A, you must be competent in all areas.

For a Grade C

You must have a reasonable grasp of Biology but you may have weaknesses in several areas and you will be unsure of some of the reasons for the Biology.

- Many Grade C candidates are just as good at answering questions as the Grade A students but holes and weaknesses often show up in just some topics.
- To improve, you will need to master your weaknesses and you must prepare thoroughly for the exam. You must become a better all-rounder.

For a Grade E

You cannot afford to miss the easy marks. Even if you find Biology difficult to understand and would be happy with a Grade E, there are plenty of questions in which you can gain marks.

- You must memorise all definitions.
- You must practise exam questions to give yourself confidence that you do know some Biology. In exams, answer the parts of questions that you know first. You must not waste time on the difficult parts. You can always go back to these later.
- The areas of Biology that you find most difficult are going to be hard to score on in exams. Even in the difficult questions, there are still marks to be gained. Show your working in calculations because credit is given for a sound method. You can always gain some marks if you get part of the way towards the solution.

What marks do you need?

The table below shows how your average mark is transferred into a grade.

average	80%	70%	60%	50%	40%
grade	A	B	C	D	E

The A* grade

To achieve an A* grade, you will need to achieve a…
- grade A overall (80% or more on uniform mark scale) for the **whole** A level qualification.
- grade A* (90% or more on the uniform mark scale) across your A2 units.

A* grades are awarded for the A level qualification only and not for the AS qualification or individual units.

Four steps to successful revision

Step 1: Understand

- Study the topic to be learned slowly. Make sure you understand the logic or important concepts.
- Mark up the text if necessary – underline, highlight and make notes.
- Re-read each paragraph slowly.

GO TO STEP 2

Step 2: Summarise

- Now make your own revision note summary:
  What is the main idea, theme or concept to be learned?
  What are the main points? How does the logic develop?
  Ask questions: Why? How? What next?
- Use bullet points, mind maps, patterned notes.
- Link ideas with mnemonics, mind maps, crazy stories.
- Note the title and date of the revision notes
  (e.g. Biology: Homeostasis, 3rd March).
- Organise your notes carefully and keep them in a file.

This is now in *short-term memory*. You will forget 80% of it if you do not go to Step 3.
GO TO STEP 3, but first take a 10 minute break.

Step 3: Memorise

- Take 25 minute learning 'bites' with 5 minute breaks.
- After each 5 minute break test yourself:
  Cover the original revision note summary.
  Write down the main points.
  Speak out loud (record on tape).
  Tell someone else.
  Repeat many times.

The material is well on its way to *long-term memory*.
You will forget 40% if you do not do step 4. *GO TO STEP 4*

Step 4: Track/Review

- Create a Revision Diary (one A4 page per day).
- Make a revision plan for the topic, e.g. 1 day later, 1 week later, 1 month later.
- Record your revision in your Revision Diary, e.g.
  Biology: Homeostasis, 3rd March 25 minutes
  Biology: Homeostasis, 5th March 15 minutes
  Biology: Homeostasis, 3rd April 15 minutes
  ... and then at monthly intervals.

Chapter 1

Energy for life

The following topics are covered in this chapter:

- *Metabolism and ATP*
- *Respiration*
- *Autotrophic nutrition*

1.1 Metabolism and ATP

After studying this section you should be able to:

- *understand the principles of metabolic pathways*
- *understand the importance of ATP*

LEARNING SUMMARY

Metabolic pathways

AQA 4.2

Inside a living organism there are many chemical reactions occurring at the same time. They may be occurring in the same place, in different parts of the cell or in different cells. Each reaction is controlled by a different enzyme.

> **KEY POINT**
>
> All the chemical reactions occurring in an organism are called **metabolism.**

Often a number of chemical reactions are linked together. The product of one reaction acts as the substrate for the next reaction. This is called a **metabolic pathway** and each of the chemicals in the pathway are called **intermediates**.

A is the substrate for this pathway, B, C and D are intermediates and E is the product. The enzymes a, b, c and d each control a different step.

$$A \xrightarrow{a} B \xrightarrow{b} C \xrightarrow{c} D \xrightarrow{d} E$$

Metabolic reactions can be classed as one of two types. Reactions that break down complex molecules are called **catabolic reactions** or catabolism. Other reactions build up complex molecules from simple molecules. They are **anabolic reactions** or anabolism.

Anabolic reactions tend to require energy, whereas catabolic reactions release energy. The link between these two types of reactions is a molecule called **ATP**.

Adenosine triphosphate (ATP)

The breakdown of many organic molecules can release large amounts of energy. Similarly, making complex molecules such as proteins requires energy. These reactions must be coupled together. This is achieved by using **adenosine triphosphate (ATP)** molecules.

ATP is a **phosphorylated nucleotide.** (Recall the structure of DNA which consists of nucleotides.) Each nucleotide consists of an organic base, ribose sugar and phosphate group. ATP is a nucleotide with two extra phosphate groups. This is the reason for the term 'phosphorylated nucleotide'.

adenine —— ribose —— phosphate —— phosphate —— phosphate

ATP is produced from adenosine diphosphate and a phosphate group. This requires energy. The energy is trapped in the ATP molecule. An enzyme called ATPase catalyses this reaction.

			ATP synthase		
ADP	+	P	→	ATP	
adenosine diphosphate		phosphate		adenosine triphosphate	

The hydrolysis of the terminal phosphate group liberates the energy. This can then be used in a number of different ways.

ATPase is a hydrolysing enzyme so that a water molecule is needed, but this is not normally shown in the equation.

	ATPase						
ATP	→	ADP	+	P	+	energy	
adenosine triphosphate		adenosine diphosphate		phosphate			

> **KEY POINT**
>
> ATP is the cell's energy currency. A cell does not store large amounts of ATP but it uses it to transfer small packets of energy from one set of reactions to another.

Uses of ATP

- muscle contraction
- active transport
- synthesis of macromolecules
- stimulating the breakdown of substrates to make even more ATP for other uses

1.2 Autotrophic nutrition

After studying this section you should be able to:

- *describe the part played by chloroplasts in photosynthesis*
- *recall and explain the biochemical processes of photosynthesis*
- *relate the properties of chlorophyll to the absorption and action spectra*
- *understand how the law of limiting factors is linked to productivity*
- *understand that glucose can be converted into a number of useful chemicals*

LEARNING SUMMARY

Synthesising food

AQA 4.3

Autotrophic nutrition is very important. Autotrophic nutrition means that simple inorganic substances are taken in and used to synthesise organic molecules. Energy is needed to achieve this. In **photo-autotrophic nutrition** light is the energy source. In most instances, the light source is **solar energy**, the process being **photosynthesis**. Carbon dioxide and water are taken in by organisms and used to synthesise glucose, which can be broken down later during respiration to release the energy needed for life. Glucose molecules can be polymerised into starch, a storage substance. By far the greatest energy supply to support food chains and webs is obtained from photo-autotrophic nutrition. Most producers use this nutritional method.

The chloroplast

AQA ▶ 4.3

Chloroplasts are organelles in plant cells which photosynthesise. In a leaf, they are strategically positioned to absorb the maximum amount of light energy. Most are located in the palisade mesophyll of leaves, but they are also found in both spongy mesophyll and guard cells. There is a greater amount of light entering the upper surface of a leaf so the palisade tissues benefit from a greater chloroplast density.

The diagram below shows the structure of a chloroplast.

Remember that not all light reaching a leaf may hit a chloroplast. Photons of light can be reflected or even absorbed by other parts of the cell. Around 4% of light entering an ecosystem is actually utilised in photosynthesis!

Even when light reaches the green leaf, not all energy is fixed in the carbohydrate product. Just one quarter becomes chemical energy in carbohydrate.

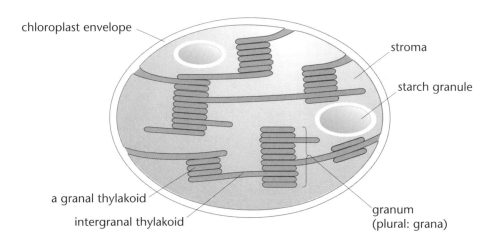

chloroplast envelope

stroma

starch granule

a granal thylakoid

intergranal thylakoid

granum
(plural: grana)

Structure and function

A system of **thylakoid membranes** is located throughout the chloroplast. These are flattened membranous vesicles which are surrounded by a liquid-based matrix, the **stroma**.

Along the thylakoid membranes are key substances:

- chlorophyll molecules
- other pigments
- enzymes
- electron acceptor proteins.

Throughout the chloroplasts, circular thylakoid membranes stack on top of each other to form **grana**. Grana are linked by longer **intergranal thylakoids**. Granal thylakoids and intergranal thylakoids have different pigments and proteins. Each type has a different role in photosynthesis.

The key substances in the thylakoids occur in specific groups comprising pigment, enzyme and electron acceptor proteins. There are two specific groups known as **photosystem I** and **photosystem II**.

Do not be confused by the photosystems. They are groups of chemicals which harness light and pass on energy. Remember this information to understand the biochemistry of photosynthesis.

The photosystems

Each photosystem contains a large number of chlorophyll molecules. As light energy is received at the chlorophyll, electrons from the chlorophyll are boosted to a higher level and energy is passed to pigment molecules known as the **reaction centre**.

> The reaction centre of photosystem I absorbs energy of wavelength 700 nanometres. The reaction centre of photosystem II absorbs energy of wavelength 680–690 nanometres. In this way, light of different wavelengths can be absorbed.
>
> **KEY POINT**

The process of photosynthesis

AQA   4.3

The process of photosynthesis is summarised by the flow diagram below.

In examinations, look out for parts of this diagram. There may be a few empty boxes where a key substance is missing. Will you be able to recall it?

- Photosynthesis harnesses solar energy.
- Photosynthesis involves light-dependent and light-independent reactions.
- Photosynthesis results in the flow of energy through an ecosystem.

Light-dependent reaction

- Light energy results in the excitation of electrons in the **chlorophyll**.
- These electrons are passed along a series of electron acceptors in the thylakoid membranes, collectively known as the **electron carrier system**.
- Energy from excited electrons funds the production of **ATP** (adenosine triphosphate).
- The final electron acceptor is **NADP$^+$**.
- Electron loss from chlorophyll causes the splitting of water (photolysis):

$$H_2O \rightarrow H^+ + OH^- \quad \text{then} \quad 4OH^- \rightarrow 2H_2O + O_2 + 4e^-$$

- Oxygen is produced, water to re-use, and electrons stream back to replace those lost in the chlorophyll.
- Hydrogen ions (H$^+$) from photolysis, together with NADP$^+$ form **NADPH$_2$**.

No ATP and NADPH$_2$ in a chloroplast would result in no glucose being made. Once supplies of ATP and NADPH$_2$ are exhausted then photosynthesis is ended. In examinations look out for the 'lights out' questions where the light-independent reaction continues for a while until stores of ATP, NADPH$_2$ and GP are used up. These questions are likely to be graph based.

Light-independent reaction

- Two useful substances are produced by the light-dependent stage, ATP and NADPH$_2$. These are needed to drive the light-independent stage.
- They react with glycerate-3-phosphate (GP) to produce a triose sugar – **triose phosphate**.
- Triose phosphate is used *either* to produce a 6C sugar *or* to form **ribulose bisphosphate** (RuBP).
- The conversion of triose phosphate (3C) to RuBP occurs in the Calvin cycle and utilises ATP, which supplies the energy required.
- A RuBP molecule (5C) together with a carbon dioxide molecule (1C) forms two GP molecules (2 × 3C) to complete the Calvin cycle.
- The GP is then available to react with ATP and NADPH$_2$ to synthesise more triose sugar or RuBP.

How do the photosystems contribute to photosynthesis?

This can be explained in terms of the **Z scheme** shown below.

The **Z scheme**, so called because the paths of electrons shown in the diagram are in a 'Z' shape.

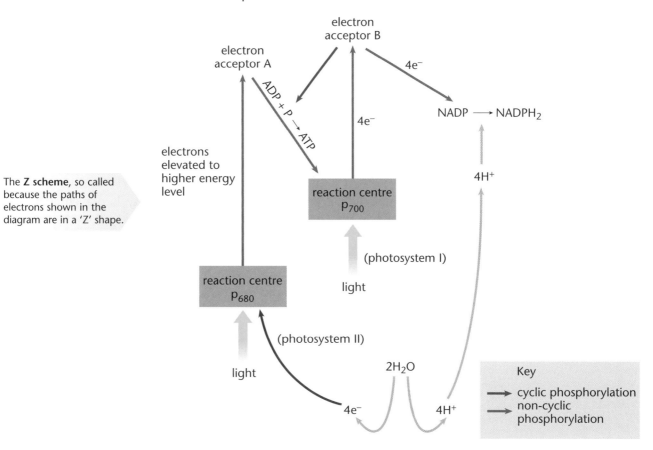

Non-cyclic photophosphorylation

- Light reaches the chlorophyll of both photosystems (P_{680} and P_{700}) which results in the excitation of electrons.
- Electron acceptors receive these electrons (**accepting** electrons is **reduction**).
- P_{680} and P_{700} have become oxidised (**loss** of electrons is **oxidation**).
- P_{680} receives electrons from the **lysis** (splitting) of water molecules and becomes neutral again (referred to as 'hydro lysis').
- Lysis of water molecules releases oxygen which is given off.
- Electrons are elevated to a higher energy level by P_{680} to electron acceptor A and are passed along a series of electron carriers to P_{700}.
- Passage along the electron carrier system funds the production of ATP.
- The electrons pass along a further chain of electron carriers to NADP, which becomes reduced, and at the same time this combines with H^+ ions to form $NADPH_2$.

After analysing this information you will be aware that in cyclic photophosphorylation P_{700} donates electrons then some are recycled back, hence 'cyclic'. In non-cyclic photophosphorylation P_{680} electrons ultimately reach NADP never to return! Neutrality of the chlorophyll of P_{680} is achieved utilising electrons donated from the splitting of water. Different electron sources hence non-cyclic.

Cyclic photophosphorylation

- Electrons from acceptor B move along an electron carrier chain to P_{700}.
- Electron passage along the electron carrier system funds the production of ATP.

Which factors affect photosynthesis?

AQA ▶ 4.3

If any process is to take place, then correct components and conditions are required. In the case of photosynthesis these are:

- light
- water
- carbon dioxide
- suitable temperature.

Additionally, it is most important that the chloroplasts have been able to develop their photosynthetic pigments in the thylakoid membranes. Without an adequate supply of magnesium and iron, a plant suffers from **chlorosis** due to chlorophyll not developing. The leaf colour becomes yellow-green and photosynthesis is reduced.

Limiting factors

If a component is in low supply then productivity is prevented from reaching maximum. In photosynthesis, **carbon dioxide** is a key limiting factor. The usual atmospheric level of carbon dioxide is 0.04%. In perfect conditions of water availability, light and temperature, this low carbon dioxide level holds back the photosynthetic potential.

Clearly **light energy** is vital to the process of photosynthesis. It is severely limiting at times of partial light conditions, e.g. dawn or dusk.

Water is vital as a photosynthetic component. It is used in many other processes and has a lesser effect as a limiting factor of photosynthesis. In times of water shortage, a plant suffers from a range of problems associated with other processes before a major effect is observed in photosynthesis.

A range of enzymes are involved in photosynthesis; therefore the process has an optimum **temperature** above and below which the rate reduces (so the temperature of the plant's environment can be limiting).

The rate of photosynthesis is limited by light intensity from points A to B. After this, a maximum rate is achieved – the graph levels off.

The rate of photosynthesis is limited by light intensity until each graph levels off. The 30°C graph shows that at 20°C, the temperature was also a limiting factor.

The lower level of CO$_2$ is also a limiting factor here. The fact that it holds back the process is shown by comparing both graph lines.

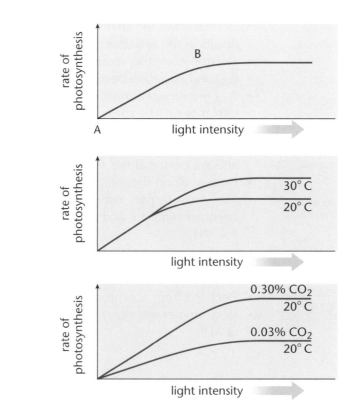

Compensation point

Another way of stating at compensation point is: 'when the rate of respiration equals the rate of photosynthesis'.

It is usual for a plant growing outside in warm conditions to have **two** compensation points every day.

Photosynthesis utilises carbon dioxide whereas respiration results in its production. At night time during darkness, a plant respires and gives out carbon dioxide. Photosynthesis only commences when light becomes available at dawn, if all other conditions are met. At one point, the amount of carbon dioxide released by respiration is totally re-used in photosynthesis. This is the **compensation point**.

Beyond this compensation point, the plant may increasingly photosynthesise as conditions of temperature and light improve. The plant at this stage still respires producing carbon dioxide in its cells and all of this carbon dioxide is utilised. However, much more carbon dioxide is needed which diffuses in from the air.

In the evening when dusk arrives, a point is reached when the rate of photosynthesis falls due to the decrease in light and the onset of darkness. The amount of carbon dioxide produced at one point is totally utilised in photosynthesis. Another compensation point has arrived.

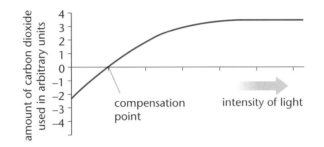

How useful is photosynthesis?

Many more substances are synthesised as a result of photosynthesis. Just a few are highlighted in this section.

Without doubt, photosynthesis is a most important process because it supplies carbohydrates and gives off oxygen. There are many more benefits in that glucose is a 'starter' chemical for the synthesis of many other substances. **Cellulose, amino acids** and **lipids** are among the large number of chemicals which can be produced as a result of the initial process of photosynthesis.

The work of the Royal Mint produces the money to run the economy; photosynthesis supplies the **energy currency** for the living world.

The table shows some examples of where and how some carbohydrates are used.

Carbohydrate	Use
deoxyribose (monosaccharide)	DNA 'backbone'
glucose (monosaccharide)	leaves, nectar, blood as energy supply
sucrose (disaccharide)	sugar beet as energy store
lactose (disaccharide)	milk as energy supply
cellulose (disaccharide)	protective cover around all plant cells
starch (polysaccharide)	energy store in plant cells
glycogen (polysaccharide)	energy store in muscle and liver

Progress check

1 In a chloroplast, where do the following take place:
   (a) light-dependent reaction
   (b) light-independent reaction?

2 (a) Which features do photosystems I and II share in a chloroplast?
   (b) Which photosystem is responsible for:
      (i) the elevation of electrons to their highest level
      (ii) acceptance of electrons from the lysis (splitting) of water?

3 (a) Complete the sentence by writing in the correct words.
      The compensation point of a plant is when the rate of ....
      equals the rate of ....

   (b) During a cloudless day in ideal conditions for photosynthesis, how many
      compensation points does a plant have? Give a reason for your answer.

4 List the three main factors which limit the rate of photosynthesis.

5 During the light-independent stage of photosynthesis, which substances are needed
   to continue the production of RuBP? Underline the substances in your answer
   which are directly supplied from the light-dependent stage of photosynthesis.

5 Triose phosphate NADPH₂, ATP and CO₂.
4 CO₂; light; temperature.
3 (a) respiration; photosynthesis (b) Two. Around dawn and dusk there will come a time when the CO₂ produced as a result of respiration is totally used up in photosynthesis.

2 (a) Each photosystem contains a large number of chlorophyll molecules. Light energy is received at the chlorophyll where electrons are boosted to a higher level. Energy is passed to pigment molecules known as the **reaction centre**. The reaction centre of each photosystem absorbs energy (but of different wavelengths).
(b) (i) photosystem I (ii) photosystem II.

1 (a) thylakoid membranes (b) stroma.

1.3 Respiration

After studying this section you should be able to:

LEARNING SUMMARY

- *recall the structure of mitochondria and relate structure to function*
- *understand that respiration liberates energy from organic molecules*
- *explain the stages of glycolysis and Krebs cycle*
- *explain the stages in the hydrogen carrier system*
- *explain the differences between anaerobic and aerobic respiration*

The site of respiration

AQA   4.4

Respiration is vital to the activities of every living cell. Like photosynthesis it is a complicated metabolic pathway. The aim of respiration is to break down **respiratory substrates** such as glucose to produce **ATP**.

Respiration consists of a number of different stages. These occur in different parts of the cell. Some stages require oxygen and some do not.

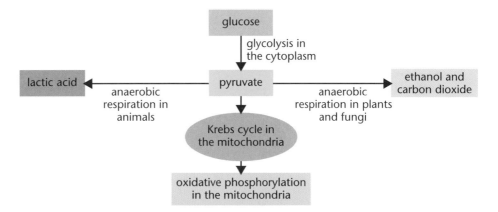

Glycolysis occurs in the cytoplasm of the cell. The pyruvate produced then enters the mitochondria. The Krebs cycle then occurs in the matrix of the mitochondria followed by oxidative phosphorylation which occurs on the inner membrane of the cristae.

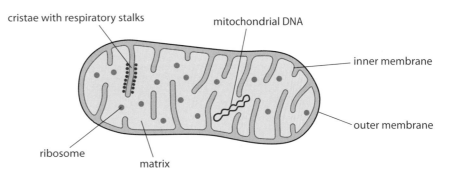

The biochemistry of respiration

AQA   4.4

Glycolysis and the Krebs cycle

Both processes produce ATP from substrates but the Krebs cycle produces **many more** ATP molecules than glycolysis. Every stage in each process is catalysed by a specific enzyme. In aerobic respiration, **both** glycolysis and the Krebs cycle are involved, whereas in anaerobic respiration only glycolysis takes place.

The flow diagram on the following page shows stages in the breakdown of glucose in glycolysis and the Krebs cycle. The flow diagram shows only the main stages of each process.

The two molecules of ATP are needed to begin the process. Each stage is catalysed by an enzyme, e.g. a decarboxylase removes CO_2 from a molecule.

The production of hydrogen atoms during the process can be monitored using DCPIP (dichlorophenol indophenol). It is a hydrogen acceptor and becomes colourless when fully reduced.

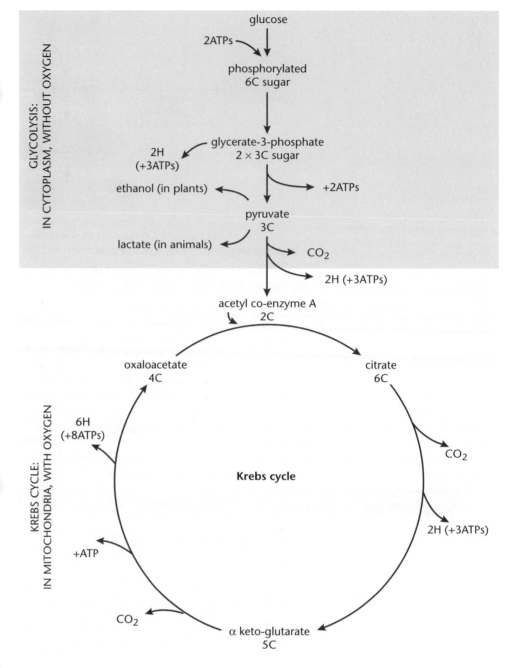

The maximum ATP yield per glucose molecule is:

GLYCOLYSIS	2
KREBS CYCLE	2
OXIDATIVE PHOSPHORYLATION	34
	= 38 ATP

The flow diagram shows that glycolysis produces 2 × 2ATP molecules but uses 2ATP so the net production is 2ATP. The Krebs cycle makes 2ATP directly. All the rest of the ATP molecules that are made (shown in brackets) are produced in **oxidative phosphorylation**.

Oxidative phosphorylation

The main feature of this process is the electron carrier or electron transport system. The hydrogen that is given off by glycolysis and the Krebs cycle is picked up by acceptor molecules such as **NAD**. These hydrogen atoms are passed along a series of carriers on the inner membrane of the mitochondrion.

Oxygen is needed at the end of the carrier chain as a hydrogen acceptor. This is why we need oxygen to live. Without it, the generation of ATP along this route would be stopped.

Oxidation	Reduction
gain of oxygen	loss of oxygen
loss of hydrogen	gain of hydrogen
loss of electrons	gain of electrons

Electron transport system

This is sometimes known as the hydrogen carrier system.

The carrier, NAD, is nicotinamide adenine dinucleotide. Similarly, FAD is flavine adenine dinucleotide.

Hydrogen is not transferred to cytochrome. Instead, the 2H atoms ionise into $2H^+ + 2e^-$. H is passed via an intermediate co-enzyme Q to cytochrome.

Only the electrons are carried via the cytochromes.

e^- is an electron.
H^+ is a hydrogen ion or proton.

An enzyme can be both an oxidoreductase and a dehydrogenase at the same time!

$2H \longrightarrow$ NAD (carrier 1)

NADH $\longrightarrow$ $+$ H^+ $\longrightarrow$ $+ATP$

H $\longrightarrow$ FAD (carrier 2)

FADH $\longrightarrow$ e^- $\longrightarrow$ $+$ H^+ $\longrightarrow$ $+ATP$

cytochrome (carrier 3)

reduced cytochrome $\longrightarrow$ e^- $+ATP$

cytochrome (carrier 4) oxidase

oxygen $\longrightarrow$ $2H^+ + O_2 + 2e^- = H_2O$

When oxidation takes place then so does reduction, simultaneously, e.g. $NADH_2$ passes H to FAD. The NAD loses hydrogen and as a result becomes oxidised. FAD gains hydrogen and becomes $FADH_2$, and is therefore reduced. The generic term for an enzyme which catalyses this is **oxidoreductase**. Additionally an enzyme which removes hydrogen from a molecule is a **dehydrogenase**. The result is that three ATPs are produced every time 2H atoms are transported.

The chemiosmotic theory

This theory also explains ATP production in photophosphorylation in the chloroplast. The only difference is that the ions are moved in the opposite direction.

It has now been shown that the carrier molecules are arranged on the membrane of the cristae in a specific way. This means that hydrogen ions are moved out of the matrix and into the space between the two membranes. This sets up a pH gradient. The hydrogen ions can re-enter the matrix through the respiratory stalks. This movement is linked to ATP production and this process is called the **chemiosmotic theory**.

Anaerobic respiration

AQA 4.4

If oxygen is in short supply then the final hydrogen acceptor for the hydrogen atoms is missing. This means that oxidative phosphorylation will stop and NAD will not be regenerated. This will result in the Krebs cycle being unable to function.

Glycolysis can continue and produce 2ATP molecules but it would soon run out of NAD as well. A small amount of NAD can be regenerated by converting the pyruvate to lactate or ethanol. This allows glycolsis to continue in the absence of oxygen. This is **anaerobic respiration**.

Ethanol is produced in plants and yeast, lactate is made in animals.

> Anaerobic respiration will make 2ATP molecules from one glucose molecule compared to a possible 38ATP in aerobic respiration.

Progress check

1 Explain how hydrogen atom production in cells during aerobic respiration results in the release of energy for cell activity.

2 Give **three** similarities between respiration and photosynthesis.

3 (a) Name the **four** carriers in the electron transport system in a mitochondrion. Give them in the correct sequence.

   (b) Name the waste product which results from the final stage of the electron transport system.

4 For each of the following statements indicate whether a molecule would be oxidised or reduced.

   (a) (i)   loss of oxygen
       (ii) gain of hydrogen
       (iii) loss of electrons

   (b) Which type of enzyme enables hydrogen to be transferred from one molecule to another?

1 Used in the electron transport system to produce ATP; 3ATP molecules produced for every 2H atoms produced; ATP → ADP + P + energy released

2 The stages of each process are catalysed by enzymes; both processes involve ATP; respiration involves GP in glycolysis and photosynthesis involves GP in the light-independent stage

3 (a) NAD → FAD → cytochrome → cytochrome → cytochrome oxidase
   (b) water

4 (a) (i) reduced (ii) reduced (iii) oxidised
   (b) Oxidoreductase

Sample question and model answer

Radioactivity is used to label molecules. They can then be tracked with a Geiger-Müller counter.

Always be ready to link the rise in one graph line with the dip of another. The relationship holds true here as substance X and RuBP are used up in the production of GP via the Calvin cycle. It is likely that some GP would have been used with substance X to make triose sugar. This is not shown on this graph.

In an experiment, pondweed was immersed in water which was saturated with radioactive carbon dioxide ($^{14}CO_2$). It was illuminated for a time so that photosynthesis took place, the light was then switched off. The graph below shows the relative levels of some substances.

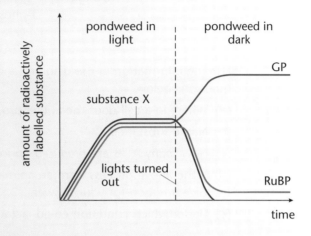

Use the graph and your knowledge to answer the following questions.

(a) (i) Substance X is produced after a substance becomes reduced during the light-dependent stage of photosynthesis. Name substance X. [1]

$NADPH_2$, reduced nicotinamide adenine dinucleotide phosphate

(ii) Explain why substance X cannot be produced without light energy. [3]

- Light energy removes electrons from chlorophyll.
- The electrons are passed along the electron carrier chain.
- The electrons are needed to reduce NADP.

(b) Explain the levels of substance X, GP and RuBP after the lights were turned off. [6]

- It seems that substance X is used to make the other two substances because it becomes used up.
- Supply of substance X cannot be produced without light energy.
- GP is made from RuBP.
- GP levels out because more $NADPH_2$ is needed to make triose sugar or RuBP, the supply being exhausted.
- RuBP levels out at a low level because more $NADPH_2$ is needed to make GP.
- ATP is needed to make RuBP, ATP is needed to make GP.

ATP is not shown on the graph. Always be ready to consider substances involved in a process but not shown. Here it is worth a mark to remember that ATP is needed to continue the light-independent system of photosynthesis.

(c) After the lights were switched off glucose was found to decrease rapidly. Explain this decrease. [1]

- Glucose is used up in respiration to release energy for the cell.

(d) Give the specific sites of each of the following stages of photosynthesis in a chloroplast: [2]
(i) light-dependent stage   thylakoid membranes
(ii) light-independent stage.   stroma

Practice examination questions

1 The flow diagram below shows stages in the process of glycolysis.

2ATPs

glucose → phosphorylated → GP → substance X → lactate
6C sugar   6C sugar   glycerate-   3C
                        3-phosphate
                       (2 × 3C)
                        2ATPs

Use the information in the diagram and your knowledge to answer the questions below.

(a) Where in a cell does the above process take place?   [1]

(b) Name substance X.   [1]

(c) How many ATPs are *produced* during the above process?   [1]

(d) Is the above process from an animal or plant?
Give a reason for your answer.   [1]

(e) Under which condition could lactate be metabolised?   [1]

[Total: 5]

2 The graph shows the relative amount of carbon dioxide taken in or evolved by a plant at different times during a day when the sun rose at 5 a.m.

(a) Explain the significance of point X.   [2]

(b) What name is given to point X?   [1]

(c) Complete the graph between 4.00 p.m. and 12.00 midnight.   [2]

[Total: 5]

3 The flow diagram below shows part of the electron carrier system in an animal cell.

FADH   →   FAD   + H^+ + e^-

(a) Where in a cell does this process take place?   [1]

(b) From which molecule did FAD receive H to become FADH?   [1]

(c) Which molecule receives the electron produced by the breakdown of FADH?   [1]

(d) As FADH becomes oxidised a useful substance is produced.
Name the substance.   [1]

[Total: 4]

Practice examination questions (continued)

4 The graph below shows the effect of increasing light intensity on the rate of photosynthesis of a plant where the concentration of carbon dioxide in the atmosphere was 0.03%.

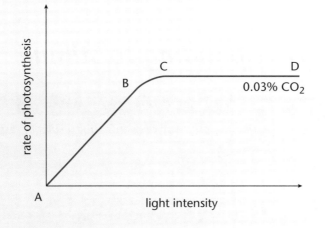

(a) Explain the effect of light intensity on the rate of photosynthesis between the following points on the graph:

(i) A and B
(ii) B and C
(iii) C and D.   [3]

(b) Draw the shape of the graph which would result from a CO_2 concentration of 0.3%.   [1]

[Total: 4]

Chapter 2
Response to stimuli

The following topics are covered in this chapter:

- Stages in responding to stimuli
- Receptors
- Plant sensitivity
- Neurone structure and function
- Response

2.1 Stages in responding to stimuli

After studying this section you should be able to:

- describe the pathway of events that results in response to stimuli

LEARNING SUMMARY

The stimulus/response pathway

AQA 5.1

The ability of plants and animals to respond to changes in their external environment is called **sensitivity**. This is a characteristic of all living organisms and is necessary for their survival. Organisms also respond to changes in their internal environment and this is covered in the next chapter.

The events involved in a response follow a similar pattern:

The responses shown by plants are often less obvious than animal responses because they are usually slower. They still involve a similar pathway of events.

In animals the communication between the receptors, the coordinating centre and the effectors is usually by neurones.

2.2 Neurone structure and function

After studying this section you should be able to:

- describe the structure of a motor neurone, a sensory neurone and a relay neurone.
- understand the function of sensory, motor and relay neurones
- understand nervous transmission by action potential
- describe the mechanisms of synaptic transmission

LEARNING SUMMARY

The structure and function of neurones

AQA 5.2

Neurones are **nerve cells** which help to coordinate the activity of an organism by transmitting **electrical impulses**. Many neurones are usually gathered together, enclosed in connective tissue to form **nerves**.

> Important features of neurones.
>
> 1 Each has a **cell body** which contains a nucleus.
> 2 Each communicates via processes **from the cell body**.
> 3 Processes that carry impulses away from the cell body are known as **axons**.
> 4 Processes that carry impulses towards the cell body are known as **dendrons**.
> 5 All neurones transmit **electrical impulses**.

KEY POINT

The nervous system consists of a range of different neurones which work in a network through the organs. The diagrams show three types of neurone.

> Notice the direction of the impulse and that motor neurones have long axons and short dendrons. This is the other way round for sensory neurones.

Key points from AS

- **The cell surface membrane**
  Revise AS pages 50–51
- **The movement of molecules in and out of cells**
  Revise AS pages 52–53
- **The specialisation of cells**
  Revise AS pages 35–36

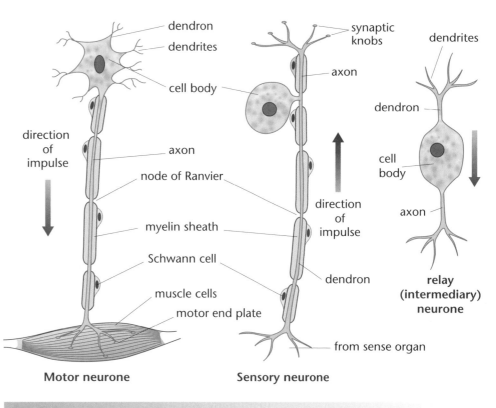

Motor neurone **Sensory neurone** **relay (intermediary) neurone**

Myelinated neurones

Sensory and motor neurones are examples of myelinated neurones. This enables them to transmit an impulse at a greater velocity. Myelinated neurones have the following characteristics:

- The axon or dendron is insulated by a **myelin sheath**.
- The myelin sheath is formed by a **Schwann cell** wrapping around the axon many times. This forms many layers of cell membrane surrounding the axon.
- At intervals there are gaps in the sheath, between each Schwann cell, called **nodes of Ranvier**.

Cross-section of axon

> The myelin sheath is often called a 'fatty' sheath because it is made of many layers of cell membrane which are composed largely of phospholipids.

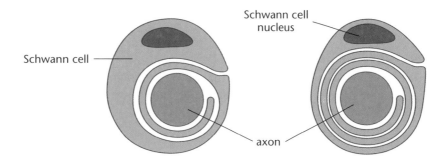

What are the roles of the sensory and motor neurones?

There are many similarities between the structure of sensory and motor neurones but they have different functions:

- The sensory neurones transmit impulses towards the central nervous system (CNS) from the receptors.
- The motor neurones transmit impulses from the CNS to effectors, such as muscles, to bring about a response.
- Relay neurones may form connections between sensory and motor neurones in the CNS.

Transmission of an action potential along a neurone

AQA 5.2

Neurones can 'transmit an electrical message' along an axon. However, you must never write this in your answers. Instead of nerve impulse, you must now use the term **action potential**.

The diagrams below show the sequence of events which take place along an axon as an action potential passes.

Resting potential

- There are 30 times more Na⁺ ions on the outside of an axon during a resting potential.
- If any Na⁺ ions diffuse in, then they are expelled by the '**sodium–potassium pump**'.
- The 'sodium–potassium pump' is an active transport mechanism by which a carrier protein, with ATP, expels Na⁺ ions against a concentration gradient and allows K⁺ ions into the axon.
- This creates a **polarisation**, i.e. there is a +ve charge on the outside of the membrane and a –ve charge on the inside.
- The potential difference is called the **resting potential** and can be measured at around –70 millivolts.

> Under resting conditions, the membrane of the axon is fairly impermeable to sodium ions.

Action potential – depolarisation

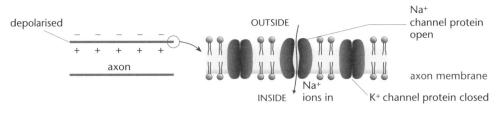

- During an action potential sodium channel proteins open to allow Na⁺ ions into the axon.
- There is now a –ve charge on the outside and a +ve charge on the inside known as **depolarisation**.
- The potential difference changes to around +50 millivolts.
- The profile of the action potential, shown by an oscilloscope, is always the same.

Action potential – repolarisation

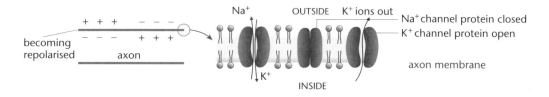

- A K$^+$ channel opens so K$^+$ ions leave the axon.
- This results in the membrane becoming polarised again.
- Any Na$^+$ ions that have entered during the action potential will be removed by the 'sodium–potassium pump'.

Measuring an action potential

- The speed and profile of action potentials can be measured with the help of an oscilloscope.
- The profile of the action potential for an organism always shows the same pattern, like the one shown.
- The changes in potential difference are tracked via a time base.
- Using the time base you can work out the speed at which action potentials pass along an axon as well as how long one lasts.

The diagram shows the typical profile of an action potential.

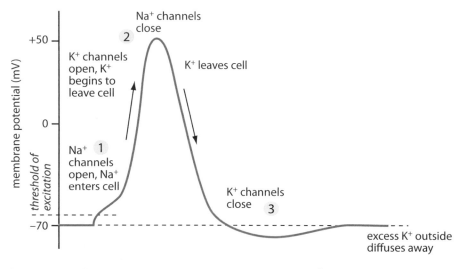

- The front of the action potential is marked by the Na$^+$ channels in the membrane opening.
- The potential difference increases to around +50 millivolts as the Na$^+$ ions stream into the axon.
- The Na$^+$ channels then close and K$^+$ channels in the membrane open.
- K$^+$ ions leave the axon and the membrane repolarises.
- During the **refractory period** no other action potential can pass along the axon, which makes each action potential separate or discrete.

Saltatory conduction

The velocity of the action potential also depends on temperature and the diameter of the axon. Wider axons and higher temperatures (up to a limit) give faster velocities.

The reason why myelinated neurones are faster than non-myelinated neurones is that the action potential 'jumps' from one node of Ranvier to the next. This is because this is the only place where Na$^+$ ions can pass across the membrane. This is called **saltatory conduction**.

Response to stimuli

Progress check

What is the function of each of the following?

(a) receptor
(b) axon
(c) myelin sheath
(d) terminal dendrites

(d) Terminal dendrites have motor end plates which can stimulate muscle tissue to contract.
(c) Myelin sheath is a membrane enclosing fat which acts as an insulator.
(b) Transmit action potential with the help of mitochondria.
(a) Receptors respond to stimulus by producing an action potential.

How do neurones communicate with each other?

AQA 5.2

The key to links between neurones are structures known as **synapses**. Terminal dendrites branch out from neurones and terminate in **synaptic knobs**. The diagram below shows a synaptic knob separated from an interlinking neurone by a synapse.

Remember an impulse can 'cross' a synapse by chemical means and the route is in ONE direction only. They cannot go back!

A synapse which conducts using acetylcholine is known as a cholinergic synapse.

There are **two** types of synapse:
• **excitatory** which can stimulate an action potential in a linked neurone
• **inhibitory** which can prevent an action potential being generated.

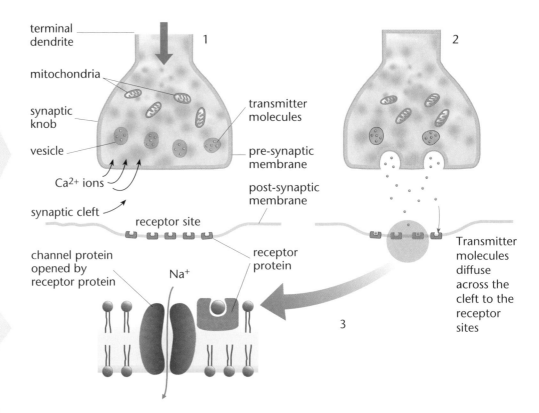

As an action potential arrives at a synaptic knob, the following sequence takes place.

• **Channel proteins in the pre-synaptic membrane** open to allow Ca^{2+} ions from the synaptic cleft into the synaptic knob.
• **Vesicles** then merge with the pre-synaptic membrane, so that **transmitter molecules** such as **acetylcholine** are **secreted** into the gap.
• The transmitter molecules diffuse across the cleft and bind with specific **sites** in **receptor proteins** in the **post-synaptic membrane**.
• Every receptor protein then opens a **channel protein** so that ions such as Na^+ pass through the post-synaptic membrane into the cell.
• The Na^+ ions **depolarise** the post-synaptic membrane.
• If enough Na^+ ions enter then depolarisation reaches a **threshold level** and an **action potential** is generated in the cell.

36

Remember that the generation of an action potential is ALL OR NOTHING. Either enough Na$^+$ ions pass through the post-synaptic membrane and an action potential is generated OR not enough reach the other side, and there is no effect.

- Enzymes in the cleft then remove the transmitter substance from the binding sites, e.g. **acetylcholine esterase** removes **acetylcholine** by hydrolysing it into choline and ethanoic acid.
- Breakdown products of transmitter substances are absorbed into the synaptic knob for re-synthesis of transmitter.

Summation

A single action potential may arrive at a synaptic knob and result in some transmitter molecules being secreted into a cleft. However, there may not be enough to cause an action potential to be generated. If a series of action potentials arrive at the synaptic knob then the build up of transmitter substances may reach the threshold and the neurone will now send an action potential. We say that the neurone has 'fired' as the action potential is produced.

KEY POINT

Progress check

1 Explain the importance of summation at a synapse.

2 The diagram shows a synaptic knob.

synaptic knob

A

post-synaptic membrane

B

(a) Name A and B

(b) Describe the events which take place after an action potential reaches a synaptic knob and a further action potential is generated as a result.

1 A single action potential may arrive at a synaptic knob; there may not be enough transmitter molecules being secreted into a cleft to cause an action potential to be generated; a series of action potentials arrive at the synapse to build up transmitter substances to reach the threshold; the neurone will now send an action potential.

2 (a) A – mitochondria; B – vesicle

(b) Ca^{2+} ions flow into the synaptic knob; transmitter molecules such as acetylcholine are secreted into the gap; the transmitter molecules bind with sites in receptor proteins in the post-synaptic membrane; this opens channel proteins so that ions such as Na$^+$ pass through the post-synaptic membrane into the cell; the post-synaptic membrane is depolarised; if *enough* Na$^+$ ions enter a threshold level is reached and an action potential is generated in the cell.

2.3 Receptors

After studying this section you should be able to:

- list the different types of receptors
- explain how receptors trigger nerve impulses
- describe how rods and cones respond to light

LEARNING SUMMARY

Types of receptors

AQA ▶ 5.1

The function of receptors is to convert the energy from different stimuli into nerve impulses in sensory neurones.

There are a range of different types of sensory cell around the body. Each type responds to different stimuli. Receptors are classified according to these different stimuli:

- **Photoreceptors**, respond to light, e.g. rods and cones in the retina.
- **Chemoreceptors**, respond to chemicals, e.g. taste buds on the tongue.
- **Thermoreceptors**, respond to temperature, e.g. skin thermoreceptors.
- **Mechanoreceptors**, respond to physical deformation, e.g. Pacinian corpuscles in the skin or hair cells in the ear.
- **Proprioreceptors**, respond to change in position in some organs, e.g. in muscles.

The more receptors there are in a position, the more sensitive it is, e.g. the fingers have many more touch receptors than the upper arm.

Stimulation of a receptor usually causes it to depolarise. This is called a **generator potential**. If this change is beyond a certain magnitude, it will trigger an action potential in a sensory neurone.

Did you know?
The umbilical cord has no receptors. It can be cut without any pain.

Some receptors are found individually in the body such as **Pacinian corpuscles** which detect pressure in the skin. Other receptors are gathered together into sense organs. An example of this is the eye which contains receptors called rods and cones.

How do the cells of the eye respond to light?

AQA ▶ 5.1

The retina in the eye contains two types of cell which are **photosensitive**, the rod cells and the cone cells. They each have different properties.

Rod cells

- They are very sensitive to the **intensity of light**, but are not sensitive to colour.
- They can respond to even dim light.
- They respond by the following reaction:
  rhodopsin → opsin + retinal
- They have low visual acuity in dim conditions.
- Opsin opens **ion channels** in the cell surface membrane which can result in the generation of an action potential.
- Rhodopsin can be re-generated during an absence of light.

400   500   600   700
wavelength/nm

The colour vision mechanism appears to require three types of cone: RED, GREEN, and BLUE and gives the *trichromatic* theory.

Cone cells

- Cone cells require **high light intensities** to be responsive – high visual acuity.
- They respond by the following reaction:
  iodopsin → photopsin + retinal
- They exist in three different types: red, green and blue, each having a different form of iodopsin:
  – **red** cones are stimulated by wavelengths of red light
  – **green** cones are stimulated by wavelengths of green light

– **blue** cones are stimulated by wavelengths of blue light

– all three cones are stimulated by white light

• Opsin again opens **ion channels** in the membranes which can lead to the generation of an action potential.

Structure of the retina

Remember that red light reaching a cone sensitive to only blue light would not stimulate the generation of an action potential. Cones are only sensitive to light of a specific range of wavelengths.

Note the direction of light as it reaches the outer retinal surface.

Visual acuity

Visual acuity is a measure of the detail we can see. The **cones** are responsible for high visual acuity (**high resolution**). **Large numbers** are packed **very close** to each other in the fovea. **ONE** cone cell synapses onto **ONE** bipolar cell which in turn synapses onto **ONE** ganglion cell as the information is relayed to the visual cortex. Spatially, much more clarity is perceived than for the rods. The image can be likened to a television picture with high numbers of pixels. Compare this with the rods. The rod cells are not packed close together so that visual acuity is low.

Did you know that the fovea consists almost entirely of cones?

What do we see with in dim conditions?

You have guessed it, rods. Their ability to detect dim light is useful but there is no colour!

Convergence

Many rods can synapse onto one bipolar cell. A ray of light reaching one rod may not be enough to stimulate an action potential along a nerve pathway. **Several** rods link to **one** bipolar cell so that enough transmitter molecules at a synapse reach the threshold level. This depolarisation results in an action potential in the bipolar cell. This is **summation**, as a result of rod cell teamwork!

2.4 Response

After studying this section you should be able to:

- describe different types of response
- understand how neurones function together in a reflex arc
- outline the features of the autonomic nervous system
- describe the structure of skeletal muscle and understand the sliding filament mechanism

LEARNING SUMMARY

Different types of response

AQA ▶ 5.1

In order to bring about a response, nerve impulses are sent to effectors via motor neurones. Some responses do not require conscious thought. These are called reflexes or reflex actions.

The reflex arc

How can we react quickly without even thinking about making a response?

Often the brain is not involved in the response so the time taken to respond to a stimulus is reduced. This rapid, automatic response is made possible by the **reflex arc**.

A reflex arc

> Small organisms have simple responses called Taxes or Kineses, which enable them to spend most of their time in a favourable environment.

Features of a reflex arc

- The stimulus elicits a response in a **receptor**.
- As a result, an **action potential** is generated along a sensory neurone.
- The **sensory neurone** enters the spinal cord via the **dorsal root** and **synapses** onto a **relay neurone**.
- This intermediate neurone synapses onto a **motor neurone** which in turn conducts the impulse to **a muscle** via its **motor end plates**.
- The muscle contracts and the arm instantly **withdraws** from the stimulus before any harm is done.
- The complete list of events takes place so quickly because the impulses do not, initially, go to the brain! The complete pathway to the muscle conducts the impulse so rapidly, before the brain receives any sensory information.

> It is other afferent neurones which *finally* take impulses to the brain enabling us to be aware of the arc which has just taken place. These afferent neurones are NOT part of the reflex arc.

> **KEY POINT**
>
> Reflexes have a high survival value because the organism is able to respond so rapidly. Additionally, they are always automatic. There are a range of different reflexes, e.g. iris/pupil reflex, and saliva production.

Autonomic nervous system

Many reflex actions are controlled by the autonomic nervous system. This is the part of the nervous system which controls our involuntary activities, e.g. the control of the heart rate. It is divided into two parts, the **sympathetic system** and the **parasympathetic system**. Each system has a major nerve from which smaller nerves branch out into key organs. In some ways, the two systems are **antagonistic** to each other but in other ways, they have specific functions not counteracted by the other. The table below shows all of the main facts for each system.

This table of features shows some key points for the autonomic system. ALERT! They are difficult to learn because of the lack of logic in the 'pattern' of functions. Take time to revise this properly because many candidates mix up the features of one system with another.

| Feature | Autonomic nervous system | |
	Sympathetic	Parasympathetic
Nerve (example)	sympathetic nerve	vagus nerve
Transmitter substance at synapses	noradrenaline	acetylcholine
Heart rate	speeds up	slows down
Iris control	dilates	constricts
Saliva	_____	flow stimulated
Gut movements	slows down	speeds up
Sweating	sweat production stimulated	_____
Erector pili muscles	contracts erector pili muscles	_____

KEY POINT

Remember that all of the above functions take place without thought. The system is truly involuntary.

One of the main functions of the autonomic nervous system is to regulate the heart rate. Under normal conditions stretch receptors called **baroreceptors** in the carotid artery, the aorta and the atria detect the pressure of the blood. If it is too high then the **cardiovascular centre** in the **medulla** of the brain will reduce the heart rate via the **vagus nerve**. The **accelerator nerve** is used to increase the heart rate if the blood pressure is too low.

During exercise, increased carbon dioxide levels are detected by **chemoreceptors**. This will result in an increase in heart rate and the blood pressure increase is ignored until the carbon dioxide levels return to normal.

Muscles as effectors

AQA ▶ 5.2/5.3

The body has a number of different effectors, but for most responses the effector is a muscle. There are three types of muscle in the body:

- skeletal/striated or voluntary muscle
- visceral/smooth or involuntary muscle
- cardiac muscle.

Smooth muscle is controlled by the autonomic nervous system, but skeletal muscle is controlled by motor neurones of the somatic nervous system. This is under voluntary or conscious control.

How do motor neurones control muscle tissue?

The link to muscle tissue is by **motor end plates** which have close proximity to the sarcoplasm of the muscle tissue. The motor end plates have a greater surface area than a synaptic knob, but their action is very similar to the synaptic transmission described on page 36. Action potentials result in muscle contraction.

No contraction would take place without the acetylcholine transmitter being released from the motor end plate. When the sarcolemma (membrane) reaches the threshold level, then the action potential is conducted throughout the sarcolemma. Contraction is initiated!

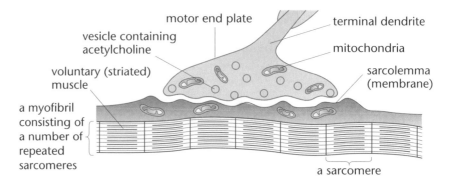

Skeletal muscle is also known as striated or striped muscle. The structure of a single muscle unit, the sarcomere, shows the striped nature of the muscle.

The sarcomere

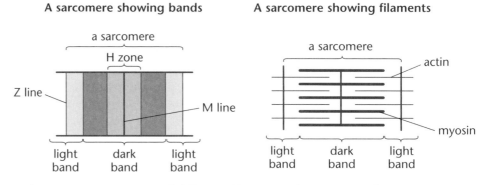

- The sarcomere consists of different filaments, thin ones (**actin**) and thick ones (**myosin**).
- These filaments form bands of different shades:
  – light band (I bands) – just actin filaments
  – dark band (A bands) – just myosin filaments or myosin plus actin.
- During contraction, the filaments slide together to form a shorter sarcomere.
- As this pattern of contraction is repeated through 1000s of sarcomeres, the whole muscle contracts.
- Actin and myosin filaments slide together because of the formation of cross bridges which alternatively build and break during contraction.
- Cross bridge formation is known as the 'ratchet mechanism'.

How does the 'ratchet mechanism' work?

To answer this question, the properties of actin and myosin need to be considered. The diagram below represents an actin filament next to a myosin filament. Many 'bulbous heads' are located along the myosin filaments (just one is shown). Each points towards an actin filament.

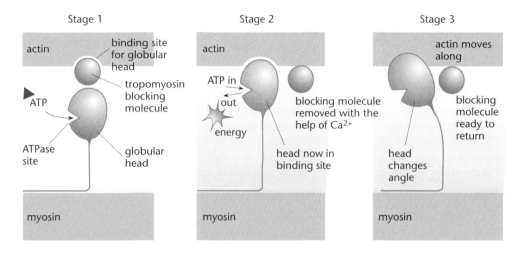

Stage 1 — actin, binding site for globular head, tropomyosin blocking molecule, ATP, ATPase site, globular head, myosin

Stage 2 — actin, ATP in / out, energy, blocking molecule removed with the help of Ca²⁺, head now in binding site, myosin

Stage 3 — actin moves along, blocking molecule ready to return, head changes angle, myosin

The sequence of the ratchet mechanism

- Once an action potential is generated in the muscle tissue then **Ca^{2+} ions** are released from the **reticulum**, a structure in the **sarcoplasm**.
- Part of the **globular head** of the myosin has an **ATPase** (enzyme) site.
- Ca^{2+} ions activate the myosin head so that this ATPase site hydrolyses an ATP molecule, **releasing energy**.
- Ca^{2+} ions also bind to **troponin** in the actin filaments, this in turn removes **blocking molecules (tropomyosin)** from the actin filament.
- This exposes the **binding sites** on the **actin** filaments.
- The globular heads of the myosin then bind to the newly exposed sites forming **actin–myosin cross bridges**.
- At the stage of energy release the myosin **heads change angle**.
- This change of angle moves the actin filaments towards the centre of each sarcomere and is termed the **power stroke**.
- More ATP binds to the myosin head, effectively causing the cross bridge to become straight and the tropomyosin molecules once again block the actin binding sites.
- The myosin is now 'cocked' and ready to repeat the above process.
- Repeated cross bridge formation and breakage results in a rowing action shortening the sarcomere as the filaments slide past each other.

Very little ATP is held in the muscles at any moment. It can be regenerated for a molecule called phosphocreatine.

actin moves this way

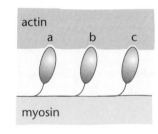

actin — a, b, c — myosin

actin — a, b, c — myosin

actin — b, c, d — myosin

Types of skeletal muscle fibres

Under the microscope skeletal muscle fibres can be divided into two types. They are slow and fast fibres. They do different jobs and so are found in different muscles.

Fibre type	Slow fibres	Fast fibres
Speed of contraction	Slow / tonic	Fast / twitch
Size of motor neurone	Small	Large
Activity used for	Responses such as maintaining posture	Quick responses such as jumping
Duration of use	Hours	Less than 5 minutes
Power produced	Low	High
Number of mitochondria	Many	Few
Found	In muscles such as the buttocks	In muscles such as the biceps
Myoglobin content	High so red	Low so white

2.5 Plant sensitivity

After studying this section you should be able to:

- understand the range of tropisms which affect plant growth
- understand how auxins, gibberellins and cytokinins control plant growth
- understand how phytochromes control the onset of flowering in plants

LEARNING SUMMARY

Plant growth regulators

AQA   5.1-2

External stimuli such as light can affect the direction of plant growth. A **tropism** is a **growth response** to an external stimulus. It is important that a plant grows in a direction which will enable it to obtain maximum supplies. Plant regulators are substances produced in minute quantities and tend to interact in their effects.

> Growth response to light is **phototropism**
> Growth response to gravity is **geotropism**
> Growth response to water is **hydrotropism**
> Growth response to contact is **thigmotropism**
> Tropisms can be positive (towards) or negative (away from).
>
> **KEY POINT**

Phototropism

This response is dependent upon the stimulus – light affecting the growth regulator, **auxin** (indoleacetic acid).

stick

auxin high
concentration here
so cells elongate

Thigmotropism helps a
climbing plant like the
runner bean to grow in
a twisting pattern
around a stick. Auxin is
redistributed away from
the contact point so the
outer cells elongate
giving a stronger outer
growth.

The diagrams show
tropic responses to light
and auxin.

Auxin and growing shoots

Auxin is produced by cells undergoing mitosis, e.g. growing tips. If a plant shoot
is illuminated from one side then the auxin is redistributed to the side furthest from
the light. This side grows more strongly, owing to the elongation of the cells,
resulting in a bend towards the light. The plant benefits from increased light for
photosynthesis. Up to a certain concentration, the degree of bending is
proportional to auxin concentration.

Tropisms in response to light from different directions

Tropism in response to auxin

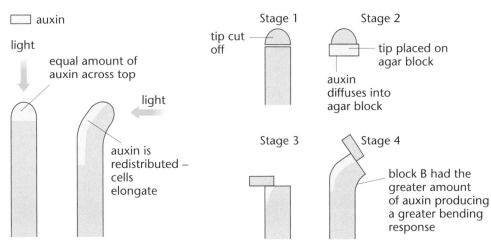

Auxin research

Many investigations of auxins have taken place using the growing tips (coleoptiles)
of grasses. Where a growing tip is removed and placed on an agar block, auxin will
diffuse into the agar. Returning the block to the mitotic area stimulates increased
cell elongation to the cells receiving a greater supply of auxin.

Is the concentration of auxin important?

It is important to consider the implications of the concentration of auxin in a tissue.
The graph below shows that at different concentrations, auxin affects the shoot
and the root in different ways.

Analyse this graph
carefully. It shows how
the same substance can
both stimulate or inhibit,
depending on
concentration.

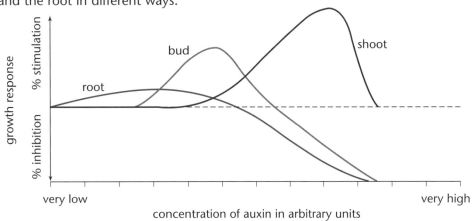

The graph shows:

- auxin has no effect on a shoot at very low concentration
- at these very low concentrations root cell elongation is stimulated
- at higher concentrations the elongation of shoot cells is stimulated
- at these higher concentrations auxin inhibits the elongation of root cells.

Auxin and root growth

The graph above shows that auxin affects root cells in a different way at different concentrations. At the root tip, auxin accumulates at a lower point because of gravity. This inhibits the lower cells from elongating. However, the higher cells at the tip have a low concentration of auxin and do elongate. The net effect is for the stronger upper cell growth to bend the root downwards. The plant therefore has more chance of obtaining more water and mineral ions.

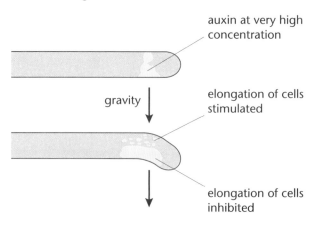

Plant growth regulators

In your examination, look out for data which will be supplied, e.g. the growth regulator gibberellin may be linked to falling starch levels in a seed endosperm and increase in maltose. Gibberellin has stimulated the enzymic activity.

Hormone	Some key functions
auxin	increased cell elongation, suppression of lateral bud development
gibberellin	cell elongation, ends dormancy in buds, promotes germination of seeds by activating hydrolytic enzymes such as amylase (food stores are mobilised!)
cytokinin	increased cell division, increased cell enlargement in leaves
ethene	promotes ripening of food

Sample question and model answer

(a) The diagram shows a neurone.

    (i) What type of neurone is shown in the diagram? [1]

        sensory neurone

    (ii) What structure would be found at A? [1]

        a receptor

    (iii) Where precisely in the body is structure C found? [1]

        in the dorsal root ganglion

    (iv) Structure B is covered by a myelin sheath.
        Explain the function of the myelin sheath. [3]

        The myelin sheath insulates the neurone;

        Ions can only pass into the neurone at the nodes of Ranvier;

        This speeds up the transmission of the nerve impulse.

(b) The diagram shows a cone from the retina.

    (i) Place an arrow on the diagram to show the direction in which light reaches the cone. [1]

    (ii) What is the function of the iodopsin in the outer segment? [4]

        when stimulated by light of the correct wavelength – breaks down to release opsin; opsin opens ion channels in the membranes; this can lead to the generation of an action potential in a bipolar cell.

    (iii) How do cones contribute to high visual acuity? [3]

        cones are tightly packed giving a high surface area; each cone synapses onto a single bipolar neurone; so the greater detail gives higher resolution.

Practice examination questions

1 The growing tips were removed from oat stems. Agar blocks containing different concentrations of synthetic auxin (IAA) replaced the tips on the oat stems. The plants were allowed to grow for a period then the angle of curvature of the stems was measured. The results are shown in the graph below.

IAA concentration in gelatin block

(a) What is the relationship between IAA concentration and curvature of the stem between points:

    (i) A and B   [1]

    (ii) C and D?   [1]

(b) Explain how IAA causes a curvature in the oat stems.   [2]

(c) Explain the effect a much higher concentration of IAA would have on the curvature of oat stems.   [2]

                                                              [Total: 6]

2 The diagram below shows a single sarcomere just before contraction.

a sarcomere

(a) Name filaments A and B.   [2]

(b) What stimulus causes the immediate contraction of a sarcomere?   [1]

(c) What happens to each type of filament during contraction?   [2]

                                                              [Total: 5]

3 The diagram below shows the profile of an action potential.

milliseconds

Explain what happens in the axon at each stage shown on the diagram.   [10]

                                                              [Total: 10]

Chapter 3

Homeostasis

The following topics are covered in this chapter:

- Hormones
- Regulation of blood sugar level
- Temperature control in a mammal
- Control of the menstrual cycle

3.1 Hormones

After studying this section you should be able to:

- define homeostasis
- describe the route of hormones from source to target organ
- understand how hormones contribute to homeostasis
- recall the roles of a range of hormones

LEARNING SUMMARY

The endocrine system

AQA ▶ 5.2/4/5

The endocrine system secretes a number of chemicals known as **hormones**. Each hormone is a substance produced by an **endocrine gland**, e.g. adrenal glands produce the hormone adrenaline. Each hormone is **transported in the blood** and has a **target organ**. Once the target organ is reached, the hormone **triggers a response** in the organ. Many hormones do this by **activating enzymes**. Others **activate genes**, e.g. steroids.

The great advantage of homeostasis is that the conditions in the environment fluctuate but conditions in the organism remain stable.

> **KEY POINT**
>
> The endocrine and nervous systems both contribute to **coordination** in animals. They help to regulate internal processes. **Homeostasis** is the maintenance of a **constant internal environment**. Nerves and hormones have key roles in the maintenance of this **steady internal state**. Levels of pH, blood glucose, oxygen, carbon dioxide and temperature all need to be controlled.

Parts of the human endocrine system (both male and female organs shown!)

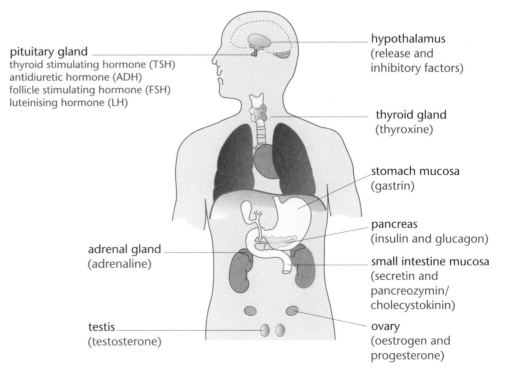

pituitary gland
thyroid stimulating hormone (TSH)
antidiuretic hormone (ADH)
follicle stimulating hormone (FSH)
luteinising hormone (LH)

hypothalamus
(release and inhibitory factors)

thyroid gland
(thyroxine)

stomach mucosa
(gastrin)

pancreas
(insulin and glucagon)

adrenal gland
(adrenaline)

small intestine mucosa
(secretin and pancreozymin/cholecystokinin)

testis
(testosterone)

ovary
(oestrogen and progesterone)

Some important mammalian hormones

The table shows the sources and some functions of a range of mammalian hormones. The pituitary gland is the key control gland. It affects many areas of the body and even stimulates, by the production of **tropic hormones**, other hormones, e.g. follicle stimulating hormone.

Hormone	Source	Effect
TSH	pituitary	stimulates the thyroid to secrete thyroxine
thyroxine	thyroid	increases metabolic rate
insulin	pancreas	reduces blood sugar
glucagon	pancreas	stimulates conversion of glycogen to glucose in the liver
ADH	pituitary	increased water reabsorption by kidney
gastrin	stomach mucosa	stimulates HCl production in stomach
secretin	intestinal mucosa	stimulates the pancreas to secrete fluid + alkali
pancreozymin	intestinal mucosa	stimulates the pancreas to secrete enzymes
FSH	pituitary	stimulates primary (Graafian) follicle to develop or testis to make sperms
LH	pituitary	stimulates ovulation or testis to make testosterone
oestrogen	ovary	stimulates development of endometrium stimulates secondary sexual characteristics
progesterone	ovary	maintains endometrium

Scientists are discovering a number of 'local hormones', such as histamine and prostaglandins. They do not travel far in the blood but only affect cells in the area.

How does a hormone trigger a cell in a target organ?

Hormones are much slower in eliciting a response than the nervous system. Rather than having an effect in milliseconds like nerves, hormones take longer. However, effects in response to hormones are often **long lasting**.

The diagram below shows one mechanism by which hormones activate target cells.

Did you know?

Each enzyme shown is constantly re-used as an active site is left free.

Look carefully at this mechanism! Just **ONE hormone molecule** arriving at the cell releases an enzyme which can be used **many** times. In turn, another enzyme is produced which can be used **many** times. One hormone molecule leads to **amplification**. This is a cascade effect.

Hormones that are proteins work in this way because they are unable to enter the cell. Steroid hormones, (e.g. oestrogen) can pass through the cell membrane as they are lipid soluble.

3.2 Temperature control in a mammal

After studying this section you should be able to:

- outline the processes which contribute to temperature regulation in a mammal
- understand how nervous and endocrine systems work together to regulate body temperature
- understand how internal processes are regulated by negative feedback

What are the advantages of controlling body temperature?

AQA ▶ 5.4

It is advantageous to maintain a constant body temperature so that the enzymes which drive the processes of life can function at an optimum level.

- **Endothermic** (warm blooded) animals can maintain their core temperature at an optimal level. This allows internal processes to be consistent. The level of activity of an endotherm is likely to fluctuate less than an ectotherm.
- **Ectothermic** (cold blooded) animals have a body temperature which fluctuates with the environmental temperature. As a result there are times when an animal may be vulnerable due to the enzyme-driven reactions being slow. When a crocodile (ectotherm) is in cold conditions, its speed of attack would be slow. When in warm conditions, the attack would be rapid.

Once the blood temperature decreases, the heat gain centre of the hypothalamus is stimulated. This leads to a rise in blood temperature which, in turn, results in the heat loss centre being stimulated. This is negative feedback! The combination of the two, in both directions, contributes to homeostasis.

The **hypothalamus** has **many functions**. It controls thirst, hunger, sleep and it stimulates the production of many hormones other than those required for temperature regulation.

How is temperature controlled in a mammal?

The key structure in homeostatic control of all body processes is the **hypothalamus**. The regulation of temperature involves thermoreceptors in the skin, body core and blood vessels supplying the brain, which link to the hypothalamus.

The diagram below shows how the peripheral nerves, hypothalamus and pituitary gland integrate nervous and endocrine glands to regulate temperature.

If there is an increase in core temperature then the hypothalamus stimulates greater heat loss by:

- vasodilation (dilation of the skin arterioles)
- relaxing of erector-pili muscles so that hairs lie flat
- more sweating
- behavioural response in humans to change to thinner clothing.

Temperature regulation model

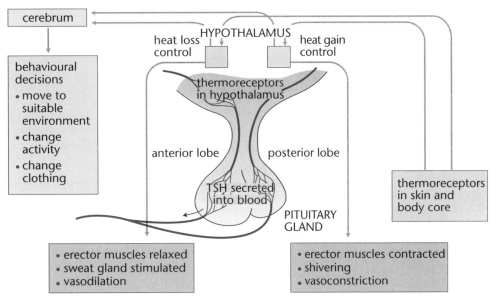

When the hypothalamus receives sensory information **heat loss** or **heat gain** control results.

A capillary bed

arteriole
(a sphincter
muscle)

venule

shunt vessel

artery   vein

A **fall** in temperature results in the following control responses.

Vasoconstriction
- Arteriole control is initiated by the hypothalamus which results in efferent neurones stimulating constriction of the arteriole sphincters of skin capillary beds.
- This deviates blood to the core of the body, so less heat energy is lost from the skin.

Contraction of the erector-pili muscles
- Erector-pili muscle contraction is initiated in the hypothalamus being controlled via efferent neurones.
- Hairs on skin stand on end and trap an insulating layer of air, so less heat energy is lost from the skin.

Sweat reduction
- The sweat glands control is also initiated in the hypothalamus, and is controlled via efferent neurones.
- Less heat energy is lost from the skin by evaporation of sweat.

Shivering
- Increased muscular contraction is accompanied by heat energy release.

Behavioural response
- A link from the hypothalamus to the cerebrum elicits this voluntary response.
- This could be to switch on the heating, put on warmer clothes, etc.

Increased metabolic rate
- The hypothalamus produces a release factor substance.
- This stimulates the anterior part of the pituitary gland to secrete TSH (thyroid stimulating hormone).
- TSH reaches the thyroid via the blood.
- Thyroid gland is stimulated to secrete thyroxine.
- Thyroxine increases respiration in the tissues increasing the body temperature.

Once a higher thyroxine level is detected in the blood the release factor in hypothalamus is inhibited so TSH release by the pituitary gland is prevented. This is **negative feedback**.

An **increase** in body temperature results in almost the **opposite** of each response described for a fall of temperature.

Vasodilation
- Arterioles of capillary beds dilate allowing more blood to skin capillary beds.

Relaxation of erector-pili muscles
- Hairs lie flat, no insulating layer of air trapped.
- More heat loss of skin.

Sweat increase
- More sweat is excreted so more heat energy from the body is needed to evaporate the sweat, so the organism cools down.

Behavioural response
- This could be to move into the shade or consume a cold drink.

Note
(a) the outline for heat loss methods does not show the nerve connections. Efferent neurones are again coordinated via the hypothalamus!
(b) heat is lost from the skin via a combination of **conduction**, **convection** and **radiation**.

Progress check

Hormone X stimulates the production of a substance in a cell of a target organ. The following statements outline events which result in the production of the substance but are in the wrong order. Write the correct order of letters.

A Hormone X is transported in the blood.
B Hormone X binds with a receptor protein in the cell surface membrane.
C The enzyme catalyses a reaction, forming a product.
D Hormone X is secreted by a gland.
E This releases an enzyme from the cell surface membrane.

D, A, B, E, C.

3.3 Regulation of blood sugar level

After studying this section you should be able to:

- *understand the control of blood glucose levels in a person*
- *describe the sites of insulin and glucagon secretion*
- *explain the functions of insulin and glucagon*

LEARNING SUMMARY

Why is it necessary to control the amount of glucose in the blood?

AQA   5.4

Glucose molecules are needed to supply energy for every living cell. The level in the blood must be high enough to meet this need (90 mg per 100 cm³ blood). This level needs to be maintained at a constant level, even though a person may or may not have eaten. High levels of glucose in the blood would cause great problems. Hypertonic blood plasma would result in water leaving the tissues by osmosis. Dehydration of organs would result in a number of symptoms.

Blood glucose regulation

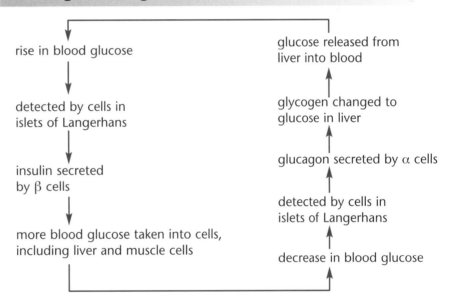

Positive feedback is rare in living organisms because any change brings about a response that increases the danger.

Negative feedback

Blood glucose regulation is an example of negative feedback. Any change in glucose level initiates changes which will result in the return of the original level – **balance is achieved**.

KEY POINT

Insulin

- Is secreted into the blood due to stimulation of pancreatic cells by a **high concentration** of glucose in the blood.
- Is produced by the β **cells** of the **islets of Langerhans** in the pancreas.
- Binds to receptor proteins in cell surface membranes activating carrier proteins to **allow glucose entry** to cells.
- Allows **excess glucose** molecules into the liver and muscles where they are converted into **glycogen** (a storage product), and some fat.

Never state that insulin changes glucose to glycogen. It allows glucose into the liver where the enzyme glycogen synthase catalyses the conversion.

Glucagon

- Is secreted into the blood due to stimulation of pancreatic cells by a **low concentration** of glucose in the blood.
- Is produced by the α **cells** of the **islets of Langerhans** in the pancreas.
- Stimulates the conversion of glycogen to **glucose**.

Diabetes

There are two types of this condition.

Type 1

- The pancreas fails to produce enough insulin.
- After a meal when blood glucose level increases dramatically, the level remains high.
- High blood glucose causes hyperglycaemia.
- Kidneys, even though they are healthy, cannot reabsorb the glucose, resulting in glucose being in the urine.
- Symptoms include dehydration, loss of weight and lethargy.

What is the answer?

- Insulin injections and a carbohydrate controlled diet.

Type 2

- This form of diabetes usually occurs in later life.
- Insulin is still produced but the receptor proteins on the cell surface membranes may not work correctly.
- Glucose uptake by the cells is erratic.
- Symptoms are similar to those for type 1 but are mild in comparison.

What is the answer?

- Dietary control including low carbohydrate intake.

More liver functions

The role of the liver in its production of bile, as well as the storage and break down of glycogen has been highlighted. The liver does so much more!

Transamination

Children need 10 essential amino acids (adults need 8). From these they can make different ones by transamination in the liver.

This is the way an R group of a keto acid is transferred to an amino acid. It replaces the existing R group with another and a new amino acid is formed.

Note the changes in the 'R' group of each acid.

Deamination

The liver has many functions including:

- detoxification of poisonous substances
- heat production.

This process is necessary to lower the level of **excess amino acids**. They are produced when proteins are digested. Nitrogenous materials have a high degree of toxicity, so the level in the blood must be limited.

$$2NH_2 - \overset{\displaystyle R}{\underset{\displaystyle H}{\overset{|}{\underset{|}{C}}}} - COOH \ + \ O_2 \ \rightarrow \ 2\overset{\displaystyle R}{\underset{\displaystyle O}{\overset{|}{\underset{||}{C}}}} - COOH \ + \ 2NH_3$$

amino acid   oxygen   keto acid   ammonia

Ornithine cycle

Both deamination and the ornithine cycle are needed to process excess amino acids. Remember that urea is a less toxic substance. The **liver makes** it but the **kidneys** help to **excrete** it.

Ammonia is immediately taken up by ornithine to help make a less toxic substance, urea.

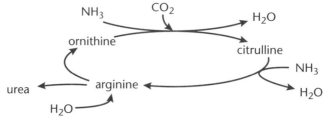

3.4 Control of the menstrual cycle

After studying this section you should be able to:

- *recall the hormones that control the menstrual cycle*
- *describe how these hormones produce cyclical changes*

LEARNING SUMMARY

What controls the menstrual cycle?

AQA 5.5

Events which take place within a female, such as ovulation and menstruation, need to be coordinated. Menstruation, the break down and loss of the lining of the uterus, would have a devastating consequence if a woman was pregnant. The foetus would be miscarried.

The menstrual cycle is controlled by the secretion of **hormones** by the **endocrine system**. The flow diagram below shows how the process achieves coordination.

Note that at the start of the menstrual cycle FSH is supported by a smaller quantity of LH to stimulate the development of a follicle. Later the proportion reverses so that a larger amount of LH is aided by a smaller amount of FSH resulting in ovulation.

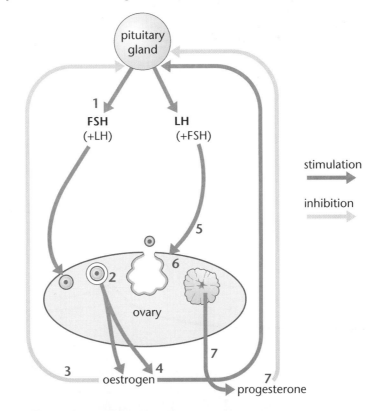

Stages of the menstrual cycle

1 (a) FSH (follicle stimulating hormone) is secreted into the bloodstream by the pituitary gland. It stimulates the development of a Graafian follicle and so triggers the development of an ovum.

   (b) At the same time a small amount of LH (luteinising hormone) is secreted by the pituitary gland which reinforces the effect of the FSH.

2 As a follicle develops, its wall (**theca**) begins to secrete oestrogen which stimulates the building of the endometrium (lining of uterus).

3 The oestrogen inhibits the secretion of FSH *temporarily* but LH secretion continues.

4 A peak of oestrogen is reached which results in a surge of LH with some FSH which is no longer inhibited.

5 When LH peaks it causes ovulation – the follicle ruptures releasing the ovum.

6 The empty follicle now changes role and becomes a corpus luteum which begins to secrete progesterone.

7 (a) Progesterone keeps the endometrium in position, as it will be needed if a foetus is to develop in the uterus.

(b) Progesterone also inhibits the secretion of any FSH or LH by the pituitary gland. Ovulation is ultimately prevented by high concentrations of progesterone.

8 If no sperm fertilises an ovum during the cycle, then the corpus luteum degenerates and a drop in progesterone takes place. Low progesterone does not inhibit the FSH and LH so that they are both able to be secreted again.

The cycle is now complete – *GO BACK TO STAGE 1!*

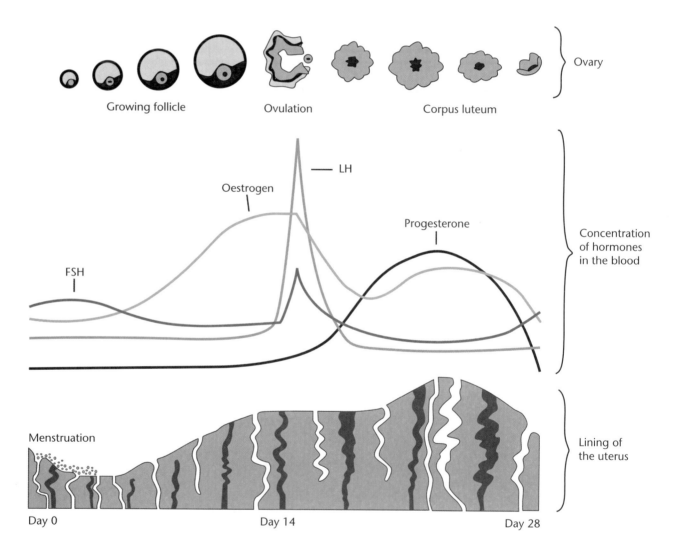

Growing follicle Ovulation Corpus luteum } Ovary

LH

Oestrogen

Progesterone

FSH

} Concentration of hormones in the blood

Menstruation

} Lining of the uterus

Day 0 Day 14 Day 28

Sample question and model answer

The graph shows changes in the concentration of progesterone in the blood during the menstrual cycle.

(a) Write down precisely where progesterone is produced. [1]

    In the corpus luteum in the ovary.

(b) How does progesterone reach the uterus? [1]

    It is carried in the plasma of the blood.

(c) On the graph, draw a line to show the concentration of oestrogen changes from day 0 to day 28. [2]

(d) Was the woman pregnant? Give a reason for your answer. [1]

    No. The levels of progesterone fall after 21 days.

(e) State one effect of progesterone. [1]

    Maintains the lining of the uterus. / Makes the muscle of the uterus contract less. / Inhibits the release of LH.

Practice examination questions

1 (a) Complete the table below to compare the nervous and endocrine systems. Put a tick in each correct box for the features shown.

	Nervous system	Endocrine system
Usually have longer lasting effects		
Have cells which secrete transmitter molecules		
Cells communicate by substances in the blood plasma		
Use chemicals which bind to receptor sites in cell surface proteins		
Involve the use of Na^+ and K^+ pumps		

[2]

(b) Name the process which keeps the human body temperature and water content of blood regulated. [1]

[Total: 3]

2 Explain the following observations.

(a) Dogs pant in hot weather. [2]

(b) The human skin looks pale in cold weather.

[2]

(c) It feels more uncomfortable in humid conditions compared to dry conditions. [2]

3 The graph below shows the relative levels of glucose in the blood of two people: A and B. One is healthy and the other one is diabetic.

(a) Which person is diabetic? Give evidence from the graph for your answer. [1]
(b) What is the evidence that both people produce insulin? [1]
(c) Where in the body is insulin produced? [2]

[Total: 4]

Chapter 4

Further genetics

The following topics are covered in this chapter:

- *Genes, alleles and protein synthesis*
- *Inheritance*

4.1 Genes, alleles and protein synthesis

After studying this section you should be able to:

- *explain how proteins are produced in cells*
- *describe various methods by which gene expression is controlled*
- *define a range of important genetic terms*

LEARNING SUMMARY

Genes, alleles and protein synthesis

AQA | 5.6

A gene is a section of DNA which controls the production of a polypeptide in an organism. The total effects of all of the genes of an organism are responsible for the characteristics of that organism. Each protein contributes to these characteristics whatever its role, e.g. structural, enzymic or hormonal.

The order of bases in the gene is called the genetic code and will code for the order of amino acids in the polypeptide. The order of amino acids is called the primary structure of the protein and will determine how the protein folds up to form the secondary and tertiary structures. The formation of a protein molecule is called protein synthesis.

Protein synthesis

The process of protein synthesis involves the DNA and several other molecules.

- **Messenger RNA:** This is a single-stranded nucleotide chain that is made in the nucleus. It carries the complementary DNA code out of the nucleus to the ribosomes in the cytoplasm.
- **mRNA polymerase:** This is the enzyme that joins the mRNA nucleotides together to form a chain.
- **ATP:** This is needed to provide the energy to make the mRNA molecule and to join the amino acids together.
- **tRNA:** This is a short length of RNA that is shaped rather like a clover leaf. There is one type of tRNA molecule for every different amino acid. The tRNA molecule has three unpaired bases that can bind with mRNA on one end and a binding site for a specific amino acid on the other end.

amino acid
binding site

messenger RNA
binding site

Key points from AS

- **The genetic code**
  Revise AS pages 73–75

The following diagrams show protein synthesis.

1 In the nucleus **RNA polymerase** links to a start code along a DNA strand.

2 RNA polymerase moves along the DNA. For every organic base it meets along the DNA a complementary base is linked to form mRNA (**messenger RNA**).

> There is no thymine in mRNA. Instead there is another base, uracil.

Pairing of organic bases				
DNA	G	C	T	A
mRNA	C	G	A	U

3 RNA polymerase links to a stop code along the DNA and finally the mRNA **moves** to a **ribosome**. The DNA stays in the nucleus for the next time it is needed.

RNA polymerase

RNA polymerase

mRNA

DNA coding strand

The transfer of the code from DNA to mRNA is called transcription.

4 Every three bases along the mRNA make up one **codon** which codes for a specific amino acid. Three complementary bases form an **anticodon** attached to one end of tRNA (**transfer RNA**). At the other end of the RNA is a specific amino acid.

5 All along the mRNA the tRNA 'partner' molecules enable each amino acid to bond to the next. A chain of amino acids (**polypeptide**) is made, ready for release into the cell.

> Note the link between each pair of amino acids along a polypeptide – the peptide link.

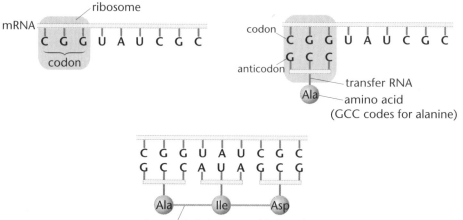

ribosome

mRNA

codon

codon

anticodon

transfer RNA

amino acid
(GCC codes for alanine)

Ala

amino acids link by peptide bonds

The conversion of the mRNA code to a sequence of amino acids is called translation.

Control of protein synthesis

AQA   5.7

In a multicellular organism every cell contains all the genetic material needed to make every protein that the organism requires. However, as they develop, cells become specialised. This means that they do not need to use every protein and so it would be a waste to make every protein all the time. Genes must be switched on and switched off.

Most of the early work on gene regulation was carried out of bacteria which are prokaryotic. During the 1950s, Jacob and Monod found that the bacterium E.Coli would only produce the enzyme lactase if lactose was present in the growth medium. The production of lactose was controlled by three different genes:

- A structural gene which codes for the enzyme.
- An operator gene which turns the structural gene on.
- A regulator gene that produces a chemical that usually stops the action of the operator gene.

If lactose is present the action of the chemical inhibitor is blocked and lactase is made.

In eukaryotic cells, gene regulation seems to be much more complicated. Cells in the early embryo are called **totipotent**. This means that they can develop into any type of cell.

These cells produce all the cells of a multicellular organism and the specialised cells have to be produced in the correct place. Scientists are trying to work out how this is done and have found genes called **homeobox genes**. These genes seem to produce proteins that act as transcription factors turning on other genes. Similar homeobox genes have been found in animals, plants and fungi.

Other factors from the cytoplasm can also affect transcription. **Steroid hormones** such as oestrogen can bind with **receptors** in the cytoplasm and then move into the nucleus causing genes to be transcripted.

Scientists have recently found a different type of RNA. This is a small double stranded molecule called siRNA (small interfering RNA). This seems to silence the action of certain genes.

> In mature plants many cells remain totipotent but in mature animals these totipotent or **stem cells** are harder to find. There is much interest in the possibility of using stem cells to cure certain degenerative diseases.

> There is much interest at present in the possible use of siRNA to treat various genetic conditions.

Essential genetic terms

AQA 5.7

It is necessary to understand the following range of specialist terms used in genetics.

Allele – an alternative form of a gene, always located on the same position along a chromosome. This position is called a locus.

> E.g. an allele coding for the white colour of petals

Dominant allele – if an organism has two different alleles then this is the one which is expressed, often represented by a capital letter.

> E.g. an allele coding for the red colour pigment of petals, **R**

Recessive allele – if an organism has two different alleles then this is the one which is **not** expressed, often represented by a lower case letter. Recessive alleles are only expressed when they are not masked by the presence of a dominant allele.

> E.g. an allele coding for the white colour pigment of petals, **r**.

Homozygous – refers to the fact that in a diploid organism both alleles for a particular gene are the same.

> E.g. **R R** or **r r**.

Heterozygous – refers to the fact that in a diploid organism both alleles for a particular gene are different.

> E.g. **R r** (petal colour would be expressed as red).

> Check out all of these genetic terms.
> - Look carefully at the technique of giving an example with each definition. Often examples help to clarify your answer and are usually accepted by the examiners.
> - In examination papers you will need to apply your understanding to **new** situations.
> - Genetics has a specialist language which you will need to use.

A number of inherited alleles of a range of genes often exhibit continuous variation, e.g. height. Each allele contributes small incremental differences. That is why there are smooth changes in height across a population.

Co-dominance – refers to the fact that occasionally two alleles for a gene are expressed equally in the organism.

E.g. **A**, **B** alleles = **AB** (blood group with antigens A and B).

Polygenic inheritance – where an inherited feature is controlled by two or more genes, along different loci along a chromosome. This results in continuous variation.

E.g. the height of a person is controlled by a number of different genes.

Remember that both sperms and ova are haploid.

Haploid – refers to a cell which has a single set of chromosomes.
E.g. a nucleus in a human sperm has 23 single chromosomes.

Diploid – refers to a cell which has two sets of chromosomes.
E.g. a nucleus in a human liver cell has 23 pairs of chromosomes.

In diploid cells one set of chromosomes is from the male parent and one from the female.

Homologous chromosomes – refers to the pairs of chromosomes seen during cell division. These chromosomes lie side by side, each gene at each locus being the same.

Often polyploid organisms cannot reproduce sexually but asexually they are successful.

Polyploid – refers to the fact that a cell has three or more sets of chromosomes. This can increase yield.

E.g. cultivated potato plants are **tetraploid**, that is four sets of chromosomes in a cell. (Tetraploidy is a form of polyploidy.)

Genotype – refers to all of the genes found in the nuclei of an organism, including both dominant and recessive alleles.

dominant
E.g.   A B c d E F g H i   (all alleles are included in a genotype)
       a B C D e f g h I
recessive

phenotype = genotype + environment

Phenotype includes all alleles which are expressed in an organism. The environment supplies resources and conditions for development. Varying conditions result in an organism developing differently. Identical twins fed different diets will show some differences, e.g. weight.

Environment has considerable effect.

Linkage – refers to two or more genes which are located on the same chromosome.

E.g. linked X-------------------Y--   not linked X-------------------- ← different
same chromosomes   -------------------Y-- ← chromosomes

Somatic cell – refers to any body cell which is not involved in reproduction.
E.g. liver cell

Autosome – refers to every chromosome apart from the sex chromosomes, X and Y.

4.2 Inheritance

After studying this section you should be able to:

- understand Mendel's laws of inheritance
- understand the principles of monohybrid inheritance
- describe the principle of sex determination
- use the Hardy–Weinberg principle to predict the numbers of future genotypes

Mendel and the laws of inheritance

AQA ▶ 4.8

Gregor Mendel was the monk who gave us our understanding of genetics. He worked with pea plants to work out genetic relationships.

> **Mendel's first law indicates that:**
> - each character of a diploid organism is controlled by a pair of alleles
> - from this pair of alleles only one can be represented in a gamete.

KEY POINT

Monohybrid inheritance

Always show your working out of a genetical relationship in a logical way, just like solving a mathematics problem.

Mendel found that when homozygous pea plants were crossed, a predictable ratio resulted. The cross below shows Mendel's principle.

pea plants   pea plants
T = TALL (dominant)   t = dwarf (recessive)

If you have to choose the symbols to explain genetics, then use something like **N** and **n**. Here the upper and lower cases are very different. **S** and **s** are corrupted as you write quickly and may be confused by the examiner awarding your marks.

A homozygous TALL plant was crossed with a homozygous recessive plant

$$TT \quad \times \quad tt$$

gametes   (T) (T)   (t) (t)
F$_1$ generation   Tt

All offspring 100% TALL, and heterozygous.

Heterozygous plants were crossed

$$Tt \quad \times \quad Tt$$

gametes   (T) (t)   (T) (t)

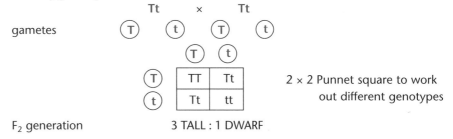

	T	t
T	TT	Tt
t	Tt	tt

2 × 2 Punnet square to work out different genotypes

In examinations you may have to work out a probability. 3:1 is the same as a 1 in 4 chance. Remember only large numbers would confirm the ratio.

F$_2$ generation   3 TALL : 1 DWARF

In making this cross Mendel investigated **one** gene only. The height differences of the plants were due to the different alleles. Mendel kept all environmental conditions the same for all seedlings as they developed. The 3:1 ratio of tall to short plants only holds true for large numbers of offspring.

Sex determination and sex linkage

AQA ▶ 4.8

The genetic information for gender is carried on specific chromosomes. In humans there are 22 pairs of autosomes plus the special sex determining pair, either **XY** (male) or **XX** (female). In some organisms such as birds this is reversed.

Some genes for sex determination are on autosomes but are activated by genes on the sex chromosomes.

Sperm can carry an X or Y chromosome, whereas an egg carries only an X chromosome.

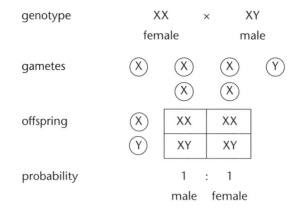

The genetic cross shown should not give you any problems at A2 Level. However, look out for the combination of another factor which will increase difficulty.

This shows how 50:50 males to females are produced.

Sex linkage

Look more closely at the structure of the X and Y chromosomes.

Why do they not look like X and Y? Only when the cells are dividing, do they take the XY shape, after chromatid formation.

Homologous part of the sex chromosomes

- Has the same genes in both sexes.
- Each gene can be represented by the same or different alleles at each locus.

Non-homologous part of the sex chromosomes

- This means that the X chromosome has genes in this area, whereas the Y chromosome, being shorter, has no corresponding genes.
- Genes in this area of the X chromosome are always expressed, because there is no potential of a dominant allele to mask them.
- There are some notable genes found on the non-homologous part, e.g. haemophilia trait, and colour blindness trait.

The sex chromosomes, X and Y, carry genes other than those involved in sex determination. Examples of such genes are:

- a gene which controls blood clotting, i.e. is responsible for the production of factor VIII vital in the clotting process
- a gene which controls the ability to detect red and green colours.

The loci of both genes are on the non-homologous part of chromosome X.

What is the effect of sex linked genes?

The fact that these genes are linked to the X chromosome has no significant effect when the **genes perform their functions correctly**. There are consequences, however, if the genes fail. This can be illustrated by a consideration of **red–green colour blindness**. When a gene is carried on a sex chromosome, the usual way to show this is by X^R.

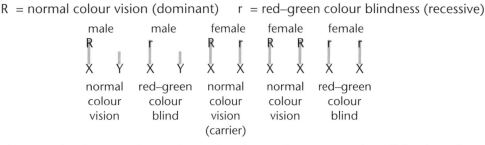

R = normal colour vision (dominant)   r = red–green colour blindness (recessive)

male	male	female	female	female
R	r	R r	R R	r r
↓	↓	↓ ↓	↓ ↓	↓ ↓
X Y	X Y	X X	X X	X X
normal colour vision	red–green colour blind	normal colour vision (carrier)	normal colour vision	red–green colour blind

The genetic diagram shows that a female needs two recessive alleles (one from each parent) to be colour blind. A male has only one gene at this locus, so one recessive allele is enough to give colour blindness. The colour blindness gene is rare, so the chances of being a colour blind female are very low.

Consider these crosses.

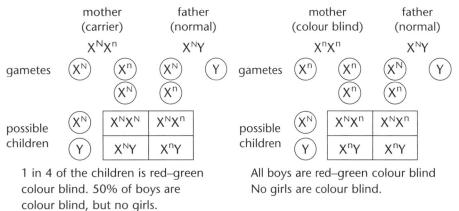

	mother (carrier)	father (normal)		mother (colour blind)	father (normal)
	$X^N X^n$	$X^N Y$		$X^n X^n$	$X^N Y$

First cross:

gametes: X^N X^n X^N Y
          X^N X^n

possible children:

	X^N	X^n
X^N	$X^N X^N$	$X^N X^n$
Y	$X^N Y$	$X^n Y$

1 in 4 of the children is red–green colour blind. 50% of boys are colour blind, but no girls.

Second cross:

gametes: X^n X^n X^N Y
          X^n X^n

possible children:

	X^N	Y
X^n	$X^N X^n$	$X^n Y$
X^n	$X^N X^n$	$X^n Y$

All boys are red–green colour blind
No girls are colour blind.

Progress check

What is the probability of a colour blind male and carrier female producing:
(a) a boy with normal colour vision
(b) a colour blind girl?
Show your working.

(answers, inverted) (a) 1 in 4 (b) 1 in 4

mother (carrier)	father (colour blind)
$X^N X^n$	$X^n Y$

	X^N	X^n
X^n	$X^N X^n$	$X^n X^n$
Y	$X^N Y$	$X^n Y$

Co-dominance

AQA ▸ 4.8

This term is given when each of two *different* alleles of a gene are expressed in the phenotype of an organism. In humans there are two co-dominant alleles that determine our blood groups. These alleles produce the antigens in blood which are responsible for our blood groups.

Consider these crosses

Blood group	Genotypes
A	$I^A I^A$, $I^A I^O$
B	$I^B I^B$, $I^B I^O$
AB	$I^A I^B$
O	$I^O I^O$

The allele for production of:

A antigen in blood = I^A
B antigen in blood = I^B
No antigen in blood = I^O

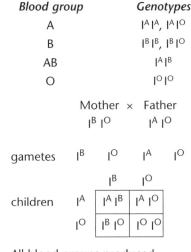

First cross:

Mother	×	Father
$I^B I^O$		$I^A I^O$

gametes: I^B I^O I^A I^O

children:

	I^B	I^O
I^A	$I^A I^B$	$I^A I^O$
I^O	$I^B I^O$	$I^O I^O$

All blood groups produced by this cross.

Second cross:

Mother	×	Father
$I^A I^B$		$I^A I^O$

gametes: I^A I^B I^A I^O

children:

	I^A	I^B
I^A	$I^A I^A$	$I^A I^B$
I^O	$I^A I^O$	$I^B I^O$

There must be a I^O from both parents to produce an O blood group.

65

In this instance, there are two co-dominant alleles, I^A and I^B. When inherited together they are both expressed in the phenotype. Group O blood does not have any antigen. I^A and I^B are both dominant over I^O.

The gene that controls our blood group is also said to show multiple alleles. This means that there are more than two possible alleles that can be found in the gene (I^A, I^O and I^B).

Remember, however, that one person can only have two copies of the alleles.

Hardy–Weinberg principle

AQA   4.8

The application of this principle allows us to **predict numbers of expected genotypes** in a population in the future. The principle tracks the proportion of two different alleles in the population.

Before applying the Hardy–Weinberg principle, the following criteria must be satisfied.

* There must be no immigration and no emigration.
* There must be no mutations.
* There must be no selection (natural or artificial).
* There must be true random mating.
* All genotypes must be equally fertile.

Once the above criteria are satisfied then **gene frequencies remain constant**.

A **gene pool** consists of all genes and their alleles, which are part of the reproductive cells of an organism. Only genes that are in cells that **can be passed on** are part of the gene pool.

Hardy–Weinberg principle: the terms identified

p   = the frequency of the dominant allele in the population

q   = the frequency of the recessive allele in the population

p^2 = the frequency of homozygous dominant individuals

q^2 = the frequency of homozygous recessive individuals

2pq = the frequency of heterozygous individuals

> **KEY POINT**
> The principle is based on two equations:
> (i) p + q = 1 (gene pool)
> (ii) $p^2 + 2pq + q^2 = 1$ (total population)

Applying the Hardy–Weinberg principle

A population of *Cepaea nemoralis* (land snail) lived in a field. In a survey there were 1400 pink-shelled snails and 600 were yellow. There were two alleles for shell colour.

Y = pink shell (dominant) y = yellow shell (recessive). Snails with pink shells can be YY or Yy. Snails with yellow shells can be yy only.

phenotype   pink   yellow
genotype   YY   Yy   yy

This part of the calculation is to find the frequency of the recessive and dominant alleles in the population.

$$q^2 = \frac{600}{2000}$$

$$= 0.3$$

$$q = \sqrt{0.30} = 0.55$$

But:
$$p + q = 1$$
$$p = 1 - 0.55$$
$$= 0.45$$

So:
$$p^2 = 0.20$$

This part of the calculation is to find the frequency of homozygous and heterozygous snails in the population.

But:
$$p^2 + 2pq + q^2 = 1$$
$$0.20 + 0.50 + 0.30 = 1$$
$$\text{YY} \quad \text{Yy} \quad \text{yy}$$

Always use the p + q = 1 equation to calculate the frequency of alleles if you are given suitable data, e.g. 'out of 400 diploid organisms in a population there were 40 homozygous recessive individuals'. 40 organisms have 80 recessive alleles.

$$q = \frac{80}{800}$$

$$= 0.1$$

From this figure you can calculate the others.

Points to note

- These proportions can be applied to the snail populations in say, 10 years in the future.
- If there were 24 000 snails in the population, then the relative numbers would be:

  YY $0.2 \times 24\,000 = 4800$

  Yy $0.5 \times 24\,000 = 12\,000$

  yy $0.3 \times 24\,000 = 7200$

- Remember that the five criteria must be satisfied if the relationship is to hold true.
- It is not possible to see which snails are homozygous dominant and which are heterozygous. They all look the same, pink! Hardy–Weinberg informs us, statistically, of those proportions.

It is also possible to apply the Hardy–Weinberg principle to a co-dominant pair of alleles. P and q are calculated by exactly the same method.

Sample question and model answer

(a) Explain the difference between sex linkage and autosomal linkage. [2]

sex linkage – genes are located on a sex chromosome

autosomal linkage – genes are located on one of the other 44 chromosomes

(b) The diagram below shows part of a family tree where some of the people have haemophilia.

This type of question is a challenge! Note the key for the symbols and then apply them to the family tree. Think logically and work up and down the diagram. In your 'live' examination write on the diagram to help you work out each individual genotype asked in the question. If there are a range of possible genotypes they may be helpful.

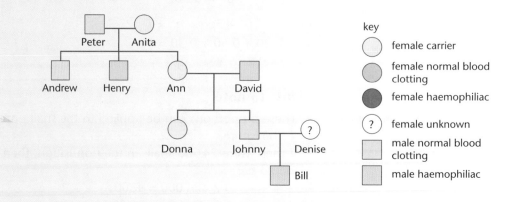

key

○ female carrier

◐ female normal blood clotting

● female haemophiliac

? female unknown

□ male normal blood clotting

□ male haemophiliac

Show the possible genotypes of Denise. Give evidence from the genetic diagram to support your answer. [3]

Let H = normal blood clotting

Let h = haemophiliac trait

The genotype can be $X^H X^h$ or $X^h X^h$

Reason – Johnny is $X^H Y$ so he is responsible for Bill's Y chromosome (Y chromosomes do not carry a blood clotting gene)

Working backwards, Bill is haemophiliac so Denise must have at least one X^h

She can, therefore, be $X^h X^h$ or $X^H X^h$

(c) Peter and Anita had three children. Andrew was born first, then Henry and finally Ann. Use the information in the diagram to answer the questions.

(i) When could genetic counselling have been given to help Peter and Anita? [1]

After the birth of Henry.

(ii) Explain the useful information which they could have been given. [3]

Since Henry is haemophiliac his genotype is $X^h Y$.

His father, Peter, has normal clotting blood so is $X^H Y$ and passes on a Y to his son, Henry.

His mother is not haemophiliac but must be a carrier, $X^H X^h$ because mother passes on X^h.

We can predict 2 in 4 children will have normal clotting of blood, 1 in 4 will be female and a carrier and 1 in 4 will be haemophiliac male.

Practice examination questions

1 (a) List the criteria which must be satisfied before applying the Hardy–Weinberg principle. [4]

(b) In a population of 160 small mammals, some had a dark brown coat and the others had a light brown coat. Dark brown (B) is dominant over light brown (b). In the population there were 48 light brown individuals. Using the Hardy–Weinberg equations calculate:

(i) the frequency of homozygous dominant and heterozygous individuals in the population [3]

(ii) how many of each of the genotypes (BB, Bb, bb) there would be in a future population of 10 000 individuals. [2]

[Total: 9]

2 Match each term with its correct definition.

A co-dominance
B polygenic inheritance
C genotype
D polyploid
E somatic

(i) a cell which is not involved in reproduction [1]

(ii) a nucleus which has three or more sets of chromosomes [1]

(iii) a feature which is controlled by two or more genes, along different loci along a chromosome [1]

(iv) two alleles which are equally expressed in the organism [1]

(v) all of the genes found in a nucleus, including both dominant and recessive alleles [1]

[Total: 5]

3 The letters below represent the organic bases along the coding strand of a DNA molecule.

CCG ATT CGA TAG

(a) What term is given to each group of three bases? [1]

(b) Give **two** functions of a group of three organic bases. [2]

[Total: 3]

Chapter 5

Variation and selection

The following topics are covered in this chapter:

- Variation
- Selection and speciation

5.1 Variation

After studying this section you should be able to:

- explain the different sources of variation in organisms
- describe different types of mutation

LEARNING SUMMARY

Variation and mutations

AQA ▶ 5.6

Meiosis and sexual reproduction can produce variation in a number of ways. These include:

- segregation or independent assortment of homologous chromosomes
- chiasmata formation leading to crossing over
- random fusion of gametes.

All these processes will combine alleles in different combinations.

The environment will also contribute to variation. The combination of environmental variation and a number of genes controlling a characteristic (polygenic inheritance) will often result in a wide range of phenotypes and continuous variation. However, the only way that new alleles can be made is by mutation.

Mutation is a change in the DNA of a cell. If the cell affected by mutation is a **somatic cell**, then its effect is **restricted** to the organism itself. If, however, the mutation affects **gametes**, then the genetic change will be inherited by the future population.

Gene mutations

A gene mutation involves a change in a single gene. This is often a point mutation.

Bases can change along DNA and this may cause mutation. One changed base along the coding strand of DNA may have a sequential effect of changing most amino acids along a polypeptide.

before mutation
TTA CCG GCC ATC

after mutation
ATT ACC GGC CAT C
This is addition!

> DNA codes for the sequence of amino acids along polypeptides and ultimately the characteristics of an organism. Each amino acid is coded for by a triplet of bases along the coding strand of DNA, e.g. TTA codes for threonine. The change in a triplet base code can result in a new amino acid, e.g. ATT codes for serine. This type of DNA change along a chromosome is known as a **point mutation**. A point mutation involves a change in a single base along a chromosome by **addition**, **insertion**, **deletion** or **inversion**.
>
> *KEY POINT*

An example of a gene mutation is a change in the DNA coding for the protein haemoglobin. This can cause sickle-cell anaemia.

Sometimes a point mutation may not cause a change in the phenotype. This is because the genetic code is degenerate. Often one amino acid has more than one triplet coding for it. Therefore a change in a base may not change the amino acid.

Key points from AS

- **Variation**
  Revise AS pages 83–84

Chromosome mutations

If a complete chromosome is added or deleted, this is a **chromosomal mutation**. Sometimes something goes wrong during meiosis and both members of a homologous pair of chromosomes move to the same pole. This produces a gamete with an extra chromosome and, after fertilisation, the zygote has an extra chromosome. This is called aneuploidy. A example is Down's syndrome where a person has an additional chromosome, totalling 47 in each nucleus rather than the usual 46.

If the spindle fails altogether, then an individual can be produced with whole extra sets of chromosomes. This is called polyploidy and is important in plant evolution.

What causes mutations?

All organisms tend to mutate randomly, so different sections of DNA can appear to alter by chance. The appearance of such a random mutation is usually very rare, typically one mutation in many thousands of individuals in a population. The rate can be increased by **mutagens** such as:

- **Ionising radiation** – including ultra violet light, X-rays and α (alpha), β (beta) and γ (gamma) rays and neutrons. These forms of radiation tend to dislodge the electrons of atoms and so disrupt the bonding of the DNA which may re-bond in different combinations.
- **Chemicals** – including asbestos, tobacco, nitrous oxide, mustard gas and many substances used in industrial processes such as vinyl chloride. Many pesticides are suspected mutagens. Dichlorvos, an insecticide, is a proven mutagen. Additionally, colchicine is a chemical derived from the Autumn crocus, *Colchicium*, which stimulates the development of extra sets of chromosomes.

Are mutations harmful or helpful?

An individual mutation may be either harmful or helpful. When tobacco is smoked, it can increase the rate of mutation in some somatic cells. The DNA disruption can result in the formation of a cell which divides uncontrollably and causes the disruption of normal body processes. This is **cancer**, and can be lethal. The presence of certain genes called **oncogenes** is thought to increase the rate of cell division and lead to cancer.

Chrysanthemum plants have a high rate of mutation. A chrysanthemum grower will often see a new colour flower on a plant, e.g. a plant with red flowers could develop a side shoot which has a different colour, such as bronze. Most modern chrysanthemums appeared in this way, production being by asexual techniques.

Some mutated human genes have, through evolution, been successful. Many successful mutations contributed to the size of the cerebrum which is proportionally greater in humans than in other primates.

5.2 Selection and speciation

After studying this section you should be able to:

- *understand the process of natural selection*
- *predict population changes in terms of selective pressures*
- *understand a range of isolating mechanisms and how a new species can be formed*
- *understand the difference between allopatric and sympatric speciation*

LEARNING SUMMARY

Natural selection

AQA ▷ 4.8

Throughout the biosphere, communities of organisms interact in a range of ecosystems. Darwin travelled across the world in his ship, the *Beagle*, observing organisms in their habitats. In 1858 Darwin published *On the Origin of Species*. In this book he gave his theory of **natural selection**.

The key features of this theory are that as organisms interact with their environment:

- individual organisms of populations are not identical, and can **vary in both genotypes and phenotypes**
- **some organisms survive** in their environment other organisms **die** before reproducing, effectively being **deleted from the gene pool**
- surviving organisms **go on to breed** and **pass on their genes** to their offspring
- this **increases the frequency of the advantageous genes** in the population.

> Learn this theory carefully then apply it to the scenarios given in your examination. Candidates often identify that some organisms die and others survive, but few go on to predict the inheritance of advantageous genes and the consequence to the species.

Consider these factors

- Adverse conditions in the environment could make a species extinct, but a range of genotypes increases the chances of the species surviving.
- Different genotypes may be suited to a changing environment, say, as a result of global warming.
- A variant of different genotype, previously low in numbers, may thrive in a changed environment and increase in numbers.
- Where organisms are well suited to their environment they have adaptations which give this advantage.
- If other organisms have been selected against, then more resources are available for survivors.
- Breeding usually produces many more offspring than the mere replacement of parents.
- Resources are limited so that competition for food, shelter and breeding areas takes place. Only the fittest survive!

What is selective pressure?

Selective pressure is the term given to a factor which has a direct effect on the numbers of individuals in a population of organisms, for example:

> 'It is late summer and the days without rainfall have caused the grassland to be parched. There is little food this year.'

> In this example, the fact that the numbers of herbivores decrease is *another* selective pressure. This time numbers of predators may decrease.

Here the **selective pressure** is a **lack of food** for the herbivores. Species which are **best adapted** to this habitat **compete** well for the limited resources and go on to survive. Within a species there is a further application of the selective pressure as weaker organisms perish and the strongest survive.

Mutations are random

New genes can appear in a species for the first time, due to a form of mutation. Over thousands of years, repeated natural selection takes place, resulting in superb adaptations to the environment.

- The Venus fly trap with its intricate leaf structures captures insects. The insects decompose, supplying minerals to the mineral deficient soil.
- Crown Imperial lilies (*Fritillaria*) produce colourful flowers, and a scent of stinking, decomposing flesh. Flies are attracted and help pollination.
- The bee orchid flower is so like a queen bee that a male will attempt mating.

Considering these examples, it is no wonder that candidates seem to consider that the organisms actively adapt to develop in these ways. They suggest that the organisms themselves have control to make active changes. **This is not so! There is no control, no active adaptation.**

> *New genes appear by CHANCE!*
>
> **KEY POINT**

Selective pressures and populations

To find out more about the effects that selective pressures can have, the **normal distribution** must be considered. The distribution below is illustrated with an example.

The mean value is at the peak. There are fewer tall and short individuals in this example. A taller plant intercepts light better than a shorter one.

normal distribution

number of individuals in population

feature, e.g. height

The further distributions below show effects of selective pressures (shown by the blue arrows). Each is illustrated with an example.

Selective pressure at both ends of the distribution causes the extreme genotypes to die. This maintains the distribution around the mean value. Mean wing length is better for flight, better for prey capture.

Selective pressure results in death of slower animals. Many die out due to predators. Faster ones (with longer legs) pass on advantageous genes. Distribution moves to the right as the average individual is now faster.

Selective pressure results in the death of organisms around the mean value. In time this can lead to two distributions. Long fur is adapted to a cold temperature and short fur to a warm temperature. The mean is suited to neither extreme.

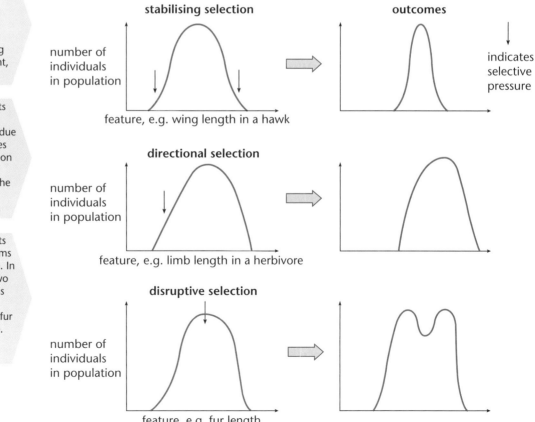

stabilising selection

number of individuals in population

feature, e.g. wing length in a hawk

outcomes

↓ indicates selective pressure

directional selection

number of individuals in population

feature, e.g. limb length in a herbivore

disruptive selection

number of individuals in population

feature, e.g. fur length

Speciation

AQA   4.8

The previous example of disruptive selection showed how two extreme genotypes can be selected. Continued selection against individuals around the former mean genotype finally results in two discrete distributions. This division into two groups may be followed by, for example, advantageous mutations. There is a probability that, in time, the two groups will become incompatible and unable to breed successfully. They have become a new species. The **development of new species is called speciation**.

To enable enough genetic differences to build up between the two groups, they must be isolated to stop them breeding. This can happen in a number of ways.

Geographical isolation

This will help you. Different finches evolved on different islands, but they did have a common ancestor.

This takes place when a population becomes divided as a result of a physical barrier appearing. For example, a land mass may become divided by a natural disaster like an earthquake or a rise in sea level. Geographical isolation followed by mutations can result in the formation of new species. This can be illustrated with the finches of the Galapagos islands. There are many different species in the Galapagos islands, ultimately from a common ancestral species. Clearly new species do form after many years of geographical isolation. This is **allopatric speciation**.

Reproductive isolation

A new pheromone is produced by several antelopes as a result of a mutation. The mainstream individuals refuse to mate as a result of this scent. An isolated few do mate. This is reproductive isolation.

This is a type of genetic isolation. Here the formation of a new species can take place in the same geographical area, e.g. mutation(s) may result in reproductive incompatibility. A new gene producing, for example, a hormone, may lead an animal to be rejected from the mainstream group, but breeding may be possible within its own group of variants. The production of a new species by this mechanism is known as **sympatric speciation**.

Sample question and model answer

The graphs below show the height of two pure breeding varieties of pea plant, Sutton First and Cava Late.

Continuous variation can confuse you sometimes when examiners display the data in categories as histograms. **This is not discontinuous!**

(a) (i) Which types of variation are shown by the pea variety, Sutton First? Give evidence from the bar graph to support your answer. [4]

Continuous variation – this is shown by the increase across the distribution (even though the peas are pure breeding).

Environmental variation – shown by the range of different heights.

(ii) Which type of variation is shown **between** varieties Sutton First and Cava Late? Give evidence from the bar graphs to support your answer. [2]

Discontinuous variation – the two distributions are separate and do not intersect.

(iii) Both Sutton First and Cava Late have compatible pollen for cross-breeding. Suggest why they do **not** cross breed. [1]

As implied by the names, Sutton First flowers before Cava Late, so that the flowers are not ready at the same time.

When you are asked to 'suggest', then a range of different plausible answers are usually acceptable.

(b) Plant geneticists considered that many years ago the two varieties of pea had the same ancestor.

(i) Suggest what, in the ancestor, resulted in the difference in height of the two varieties? [1]

mutation

(ii) Suggest what caused this change. [1]

radiation/random processes

(c) (i) Define polygenic inheritance. [1]

The inheritance of a feature controlled by a number of genes (not just a gene at one locus).

(ii) Which type of variation is a consequence of polygenic inheritance? [1]

continuous variation

Practice examination questions

1 The graph shows the birth weight of babies born in a London hospital between 1935 and 1946. It also shows the chance of the babies dying within two months of birth.

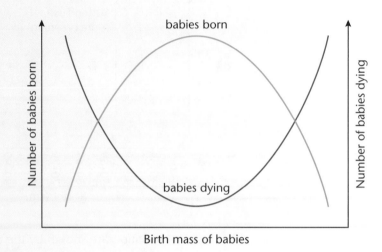

(a) What type of variation is shown by the birth weight of the babies? [1]

(b) What factors decide the birth weight of a baby? [2]

(c) Scientists argue that the information in the graphs shows that stabilising selection acts on the birth weight of babies. Explain why they think this. [3]

(d) The data was collected between 1935 and 1946. Modern medical techniques may have altered the selection pressure on the birth weight of babies. Explain why. [2]

[Total: 8]

2 (a) Explain the difference between allopatric and sympatric speciation. In each instance use an example to illustrate your answer. [6]

(b) How is it possible to find out if two female animals are from the same species? [2]

[Total: 8]

Chapter 6

Biotechnology and genes

The following topics are covered in this chapter:

- *Mapping and manipulating genes*

6.1 Mapping and manipulating genes

After studying this section you should be able to:

- describe a range of techniques used in mapping genes
- describe how genes can be transferred between a range of different organisms

Identifying genes

AQA ▶ 5.8

In order to identify the genes of an individual, a number of different processes are used.

Polymerase chain reaction

The polymerase chain reaction (PCR) is used to make numerous copies of a section of DNA. This is called **amplifying** the DNA. It uses the principle of semiconservative replication of DNA to produce new molecules that can in turn act as templates to produce more molecules. This therefore sets up a chain reaction. The enzyme DNA polymerase is used to copy the DNA.

Electrophoresis

This is used to separate sections of DNA according to their size.

Enzymes called restriction endonucleases can be used to cut up an organism's DNA.

- DNA sections are put into a well in a slab of agar gel.
- The gel and DNA are covered with buffer solution which conducts electricity.
- Electrodes apply an electrical field.
- Phosphate groups on DNA are negatively charged causing DNA to move towards the anode.
- Smaller pieces of DNA move more quickly down the agar track; larger ones move more slowly, leading to the formation of bands.

Genetic fingerprinting

PCR and electrophoresis have many applications. DNA is highly specific so the bands produced using this process can help with identification. In some crimes, DNA is left at the scene. Blood and semen both contain DNA specific to an individual. DNA evidence can be checked against samples from suspects. This is known as **genetic fingerprinting**. Genetic fingerprinting can be used in paternity disputes. Each band of the DNA of the child must correspond with a band from *either* the father or the mother.

Isolating genes

Along chromosomes are large numbers of genes. Scientists may need to identify and isolate a useful gene; one way of doing this is to use the enzyme **reverse transcriptase**. This is produced by viruses known as **retroviruses**. Reverse transcriptase has the ability to help make DNA from mRNA.

Stage 1

When a polypeptide is about to be made at a **mRNA** ribosome, reverse transcriptase allows a strand of its coding DNA to be made.

Stage 2

The single stranded DNA is parted from the mRNA.

Stage 3

The other strand of DNA is assembled using DNA polymerase.

Using this principle, the exact piece of DNA which codes for the production of a vital protein can be made.

Progress check

1   A length of DNA was prepared and then electrophoresis was used to separate the sections. The statements below describe the process of electrophoresis but they are in the wrong order. Write the letters in the correct sequence.

   A   electrodes apply an electrical field
   B   DNA sections are put into a well in a slab of agar gel
   C   smaller pieces of DNA move more quickly down the agar track with larger ones further behind
   D   the gel and DNA are then covered with buffer solution which conducts electricity
   E   restriction endonucleases can be used to cut up the DNA

2   Reverse transcriptase is an enzyme which enables the production of DNA from RNA. Work out the sequence of organic bases along the DNA of the following RNA sequence.

   A A U GCCCGGAUU

   2 RNA AAUGCCCGGAUU   DNA₂ AATGCCCGGATT   DNA₁ TTACGGGCCTAA
   1 E B D A C

The Human Genome Project

The Human Genome Project is an analysis of the complete human genetic make-up, which has mapped the organic base sequences of the nucleotides along our DNA.

A brief history

*   1977 Sanger devised DNA base sequencing.
*   1986 The Human Genome Project was initiated in the USA and the UK.
*   1996 30 000 genes were mapped.
*   1999 one billion bases were mapped including all of chromosome 22.
*   2000 chromosome 21 was mapped with the human genome almost complete.
*   2001 human genome mapping complete.

Some important points

*   The genome project will sequence the complete set of over 100 000 genes.
*   Only around 5% of the base pairs along the DNA actually result in the expression of characteristics. These DNA sequences are known as **exons**.
*   95% of DNA base sequences are not transcribed and do not appear to be involved in the expression of characteristics. These are known as **introns**.
*   Introns do not outwardly seem to be responsible for characteristics. It is likely that they may be regulatory, perhaps in multiple gene role.

Effects of single nucleotide polymorphism

Example

5 base sequences from five people →

GTATAGCCGCAT   1
GTATAGCCGCAT   1
GTATAGCCGCAT   1
GTATAGCCGCCT   2
GTATAGCCGCCT   2

Version 1 = ●
Version 2 = ●

Proportion of the SNP in healthy members of population:

Proportion of the SNP in diseased members of population:

A greater incidence of an SNP in people with a disease may point to a cause.

Single nucleotide polymorphisms (SNPs)

Around 99.9% of human DNA is the same in all individuals. Merely 0.1% is different! The different sequences in individuals can be the result of **single nucleotide polymorphism**. One base difference from one individual to another at a site may have no difference. Up to a maximum of six different codons can code for one amino acid. An SNP will not necessarily have any effect.

Some SNPs do change a protein significantly. Such changes may result in genetic disease, resistance or susceptibility to disease.

How can the mapping of SNPs be useful?

- The mapping of SNPs along chromosomes signpost where base differences exist.
- Across the gene pool a pattern of SNP positions will be evident.
- There may be a high frequency of common SNPs found in the DNA of people with a specific disease.
- This highlights interesting sites for future research and will help to find answers to genetic problems.

Benefits obtained from the Human Genome Project

Ultimately, the human genome data will be instrumental in the development of drugs to treat genetic disease. Additionally, by analysis of parental DNA, it will be possible to give the probability of the development of a specific disease or susceptibility to it, in offspring. Fetal DNA, obtained through amniocentesis or by chorionic villi sampling, will give genetic information about an individual child.

Genetic counsellors will have more information about an individual than ever before. Companies will be able to produce 'designer drugs' to alleviate the problems which originate in our DNA molecules. Soon the race will begin to produce the first crop of drugs to treat or even cure serious genetic diseases. Look to the media for progress updates.

Manipulating DNA

AQA ▶ 5.8

Scientists have developed methods of manipulating DNA. It can be transferred from one organism to another. Organisms which receive the DNA then have the ability to produce a new protein. This is one example of **genetic engineering**.

> The genetic code is universal. This means that it is possible to move genes from one organism to another and the recipient organism may be from a different species. The DNA will still code for the same protein.
>
> **KEY POINT**

Genes have now been transferred to and from many different types of organisms.

Here are some examples:

- From humans to bacteria: this technique produced the first commercially available genetically engineered product, insulin.
- From plants to plants: this technique has been used to produce GM crops such as Golden Rice that contain vitamin A.
- Into humans: this technique may be successful in treating genetic conditions such as cystic fibrosis (but it is not a cure).

Gene transfer to bacteria

The gene which produces human insulin was transferred from a human cell to a bacterium. The new microorganism is known as a **transgenic bacterium**. The process which follows shows how a human gene can be inserted into a bacterium.

The human gene for insulin is produced using an enzyme called **reverse transcriptase**. This converts the mRNA coding for insulin back into DNA. In this way all the introns are removed. Then the DNA can be inserted.

Restriction endonucleases are produced by some bacteria as a defence mechanism. They cut up the DNA of invading viruses. This can be exploited during gene transfer.

Note that **both** the donor DNA and recipient plasmid DNA are cut with the same enzyme. This allows the new gene to be a matching fit.

Many exam candidates fail to state that the plasmids are cloned inside the bacterium.

The bacteria themselves are also cloned. There may be two marks in a question for each cloning point!

1   An enzyme known as **restriction endonuclease** cuts the DNA and the gene was removed. Each time a cut was made the two ends produced were known as 'sticky ends'.

2   Circles of DNA called **plasmids** are found in bacteria.

3   A plasmid was taken from a bacterium and cut with the same restriction endonuclease.

4   The human gene was inserted into the plasmid. It was made to fix into the open plasmid by another enzyme known as **ligase**.

5   The plasmid **replicated** inside the bacterium.

6   Large numbers of the new bacteria were produced. Each was able to secrete perfect human insulin, helping diabetics all over the world.

Gene transfer to plants

Inserting genes into crop plants is becoming increasingly important in meeting the needs of a rising world population. One example of this is the production of a type of rice called Golden Rice. This contains a gene that produces vitamin A. The aim is to prevent vitamin A deficiency which can lead to blindness.

In **plants** there is an important technique which uses a **vector** to insert a novel gene. The vector is the bacterium *Agrobacterium tumefaciens*.

Agrobacterium tumefaciens
- This is a **pathogenic bacterium** which **invades** plants forming a gall (abnormal growth).
- The bacterium contains **plasmids** (circles of DNA) which carry a gene that stimulates tumour formation in the plants it attacks.
- The part of the plasmid which does this is known as the **T-DNA region** and can insert into any of the chromosomes of a host plant cell.
- Part of the T-DNA controls the production of two growth hormones, auxin and cytokinin.
- The extra quantities of these hormones stimulate rapid cell division, the cause of the tumour.

KEY POINT

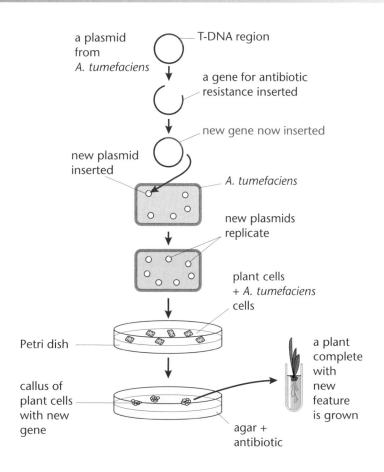

How can *Agrobacterium tumefaciens* be used in gene transfer?

The principle of using *A. tumefaciens* can be used in gene transfer to many different plants. Applications are at an early stage of development.

KEY POINT

- Firstly, the DNA section controlling auxin and cytokinin was deleted, tumours were not formed, and cells of the plant retained their normal characteristics.
- A gene which gave the bacterial cell **resistance to a specific antibiotic** was inserted into the T-DNA position.
- The **useful gene** (e.g. the gene for vitamin A) was **inserted into a plasmid.**
- Plant cells, minus cell walls, were removed and put into a Petri dish with nutrients and *A. tumefaciens*, which contained the engineered plasmids.
- The cells were **incubated** for several days, then transferred to another Petri dish containing nutrients plus the specific antibiotic.
- **Only plant cells with antibiotic resistance and the desired gene grew.**
- Any surviving cells grew into a callus, from which an adult plant formed, complete with the transferred gene.

Inserting genes into humans

The idea of changing a person's genes in order to cure genetic disease is called **gene therapy.**

There are two main possibilities:

- **Somatic cell therapy** in which the genes are inserted into the cells of the adult where they are needed.
- **Germ line gene therapy** involves changing the genes of the gametes or early embryo. This means that all the cells of the organism will contain the new gene.

Sample question and model answer

The diagrams below show the transfer of a useful gene from a donor plant cell to the production of a transgenic crop plant. The numbers on the diagram show the stages in the process.

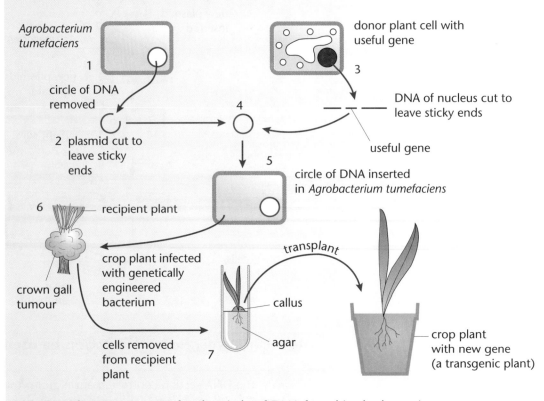

Look out for transgenic stories in the media. The principles are often the same. This will prepare you for potentially new ideas in your 'live' examinations. You could encounter the same account!

(a) Give the correct name for the circle of DNA found in the bacterium, *A. tumefaciens*. [1]

plasmid

(b) The same enzyme was used to cut the DNA of the bacterium and of the plant cell.

(i) Name the type of enzyme used to cut the DNA. [1]

restriction endonuclease

(ii) Explain why it is important to use exactly the same enzyme at this stage. [2]

The same enzyme produces the same sticky ends.

Complementary sticky ends on the donor gene bind with the sticky ends of the plasmid.

This question covers key techniques in gene transfer. Be prepared for your examination.

(iii) Which type of enzyme would be used to splice the new gene into the circle of DNA? [1]

ligase

(c) How was the new gene incorporated into the DNA of the crop plant cells? [2]

Crop plant infected by genetically engineered bacterium.

The DNA of bacterium causes a change in the DNA of the crop plant to produce the gall or tumour cells.

(d) How would you know if the gene had been transferred successfully? [1]

The feature would be expressed in the transgenic plants.

Practice examination questions

1 A new genetically modified soya bean plant has been developed. It has a new gene which prevents it from being killed by herbicide (weed killer).

(a) Describe the stages which enable a gene to be transferred from one organism to another. [5]

(b) Suggest how the genetically modified soya plants could result in higher bean yields. [3]

[Total: 8]

Chapter 7

Ecology and populations

The following topics are covered in this chapter:

- *Investigation of ecosystems*
- *Colonisation and succession*
- *Human populations*

7.1 Investigation of ecosystems

After studying this section you should be able to:

- *use the capture, mark, recapture technique to assess animal populations*
- *use quadrats to map the distribution of organisms*
- *understand the factors that affect the distribution of organisms*
- *describe conservation techniques and methods of population control*

LEARNING SUMMARY

Measurement in an ecosystem

AQA ▶ 4.1

The study of ecology investigates the inter-relationships between organisms in an area and their environment. The area in which organisms live is called a **habitat**. The combination of the organisms that live in a habitat and the physical aspects of the habitat is called an **ecosystem**.

Estimating populations

All the individuals of one species living together in a habitat are called a **population**. The size of plant populations can be estimated by using a quadrat placed at random. Animals, however, do not tend to stay still for long enough to be sampled using a quadrat. The population size of an animal species can be estimated by using capture–recapture.

Capture, mark, release, recapture

This is a method which is used to estimate animal populations. It is an appropriate method for motile animals such as shrews or woodlice. The ecologist must always ensure minimum disturbance of the organism if results are to be truly representative and that the population will behave as normal.

The technique

- Organisms are captured, *unharmed*, using a quantitative technique.
- They are counted then discretely marked in some way, e.g. a shrew can be tagged, a woodlouse can be painted (*with non-toxic paint*).
- They are released.
- Organisms from the same population are recaptured, and another count is made, to determine the number of marked animals and the number unmarked.

The calculation

Before using the technique you must be assured that:

- there is no significant migration
- there are no significant births or deaths
- marking does not have an adverse effect, e.g. the marking paint should not allow predators to see prey more easily (or vice versa)
- organisms integrate back into the population after capture.

Remember that the method is suitable for large population size only.

S = total number of individuals in the total population.
S_1 = number captured in sample one, marked and released, e.g. 8.
S_2 = total number captured in sample two, e.g. 10.
S_3 = total marked individuals captured in sample two, e.g. 2.

$$\frac{S}{S_1} = \frac{S_2}{S_3} \quad \text{so, } S = \frac{S_1 \times S_2}{S_3} \quad \text{population} = \frac{8 \times 10}{2} = 40 \text{ individuals}$$

Remember the equation carefully. You will **not** be supplied with it in the examination, but you will be given data.

Measuring the distribution of organisms

This can be measured using another quadrat technique called a belt transect. This method should be used when there is a **transition** across an area, e.g. across a pond or from high to low tide on the sea shore. Use belt transects where there is **change**. The belt transect is a line of quadrats. In each quadrat a measurement such as density can be made. One transect is not enough! Always do a number of transects then find an average for quadrats in a similar zone.

A bar graph would be used to show the **distribution** of plant species across the pond. Note that there would be more than just two species. The graphs show how you could illustrate the data. Clearly flag irises occupy a different niche to water lilies.

A simplified results table

Quadrat no.	flag iris	water lily
1	10	0
2	7	0
3	1	0
4	0	5
5	0	4
6	0	0
7	0	5
8	0	3
9	0	0
10	1	0
11	8	0
12	4	0

This is just one belt transect. A number would be used and an average taken for each corresponding quadrat.

Other uses of quadrats

Quadrats can also be used to survey animal populations. It is made easier if the organisms are **sessile** (*they do not move from place to place*), e.g. barnacles on a rock. In a pond the belt transect could be coupled with a kick sampling technique. Here rocks may be disturbed and escaping animals noted. Adding a further technique can help, such as using a catch net in the quadrat positions. The principle here is that the techniques are **quantitative**.

KEY POINT

Factors that determine population size

Graphical data can show relative numbers and distribution of organisms in a habitat. The ecologist is interested in the factors that determine the size and distribution of organisms.

These factors can be **biotic** or **abiotic**.

Abiotic factors are non-living factors. They include:

- carbon dioxide level
- oxygen level
- pH
- light intensity
- mineral ion concentration
- level of organic material.

Biotic factors are living factors.

They include:

- **Competition** This occurs when organisms are trying to get the same resources. There are two types. **Interspecific competition** takes place when **different** species share the same resources. **Intraspecific competition** takes place when the **same** species share the same resources.
- **Predation** This involves feeding relationships.

Predators and prey

There can be many examples of this type of relationship in an ecosystem. **Primary consumers** rely on the **producers**, so a flush of new vegetation may give a corresponding increase in the numbers of primary consumers. Predators which eat the primary consumers may also follow with a population increase. Each population of the ecosystem may have a **sequential effect** on other populations. Ultimately, the ecosystem is in **dynamic equilibrium** and has limits as to how many of each population can survive, i.e. its **carrying capacity**.

Note that graphs are often given in predator–prey questions. A flush of spring growth is often responsible for the increase in prey. Plant biomass may not be shown on the graph! Candidates are expected to suggest this for a mark. Also remember that as prey increase, their numbers will go down when eaten by the predator. Predator numbers rise after this.

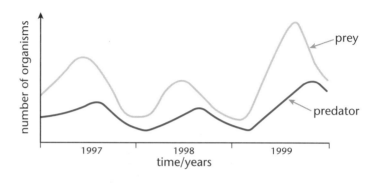

Ecological conservation

AQA 4.7

In a world where human population increase is responsible for the destruction of so many habitats, it is necessary to retain as many habitats as possible. Ecological surveys report to governments and difficult decisions are made. Fragile habitats like the bamboo woodlands of China support a variety of wildlife. Conservation areas need to be kept and maintained to prevent extinction of organisms at risk. In the UK we have **sites of special scientific interest (SSSI)** which are given government protection.

Conservation requires management

Although the word 'conservation' implies to 'keep' something as it is, much effort is needed. An area of climax vegetation, e.g. oak woodland, is less of a problem, since it will not change if merely left to its own devices. However, many of the seral stages, e.g. birch woodland along the route to climax, require much maintenance.

Animal populations need our help, especially when it is often by our own introduction that specific species have colonised an area. Deer introduced into a forest may thrive initially but due to an efficient reproductive rate exceed the carrying capacity of the habitat. **Carrying capacity** is the population of the species which can be adequately supported by the area.

Sometimes herbivores could destroy their habitat by overgrazing, and so must be **culled**. **Predators** could be introduced to reduce numbers, but they also may need culling at some stage. **Difficult decisions** need to be taken. In the aquatic habitats similar problems exist. Cod populations in the North Sea are being reduced by over-fishing. Agreements have been made by the EU to **reduce fishing quotas** and create **exclusion zones** to **allow fish stocks to recover**. Even before this agreement, smaller fish had to be returned to the sea after being caught to increase the chances of them growing to maturity and breeding successfully.

Endangered species require protection

All over the world many animals and plants are at the limits of their survival. The World Wide Fund for Nature is a charity organisation which helps. The organisation receives support from the public and artists such as David Shepherd. He gives donations from the sale of all of his wildlife paintings, helping to maintain the profile of animals so that we invest in survival projects like protected reserves.

7.2 Human populations

After studying this section you should be able to:

- *understand the growth of the human population*
- *understand population structure*

Growth of the human population

AQA ▶ 4.1

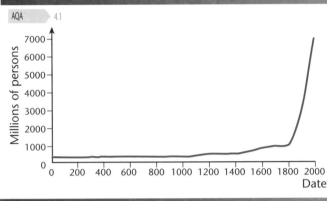

It is difficult to estimate the size of the human population thousands of years ago but most estimates produce graphs that look like the one here.

The increase was fairly slow up until about 300 years ago. Recently, a more rapid increase was brought about by improvements in medicine and healthcare, industrial productivity and food production.

There are many predictions as to how long and how fast this increase can continue.

Population structure

AQA ▶ 4.1

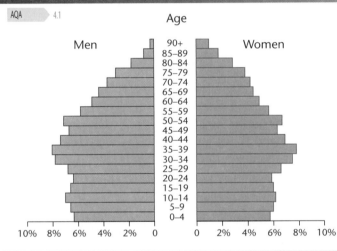

In any population there will be individuals of different ages and of different sex. A population pyramid shows how each population varies in terms of its composition.

The graph for the UK shows a number of interesting features, such as a bulge in 30 to 40 year olds. It also shows that women have a longer **life expectancy** than men.

7.3 Colonisation and succession

After studying this section you should be able to:

- *understand how colonisation is followed by changes*
- *understand how colonisation and succession lead to a climax community*

How colonisation and succession take place

AQA ▶ 4.7

Any area which has never been inhabited by organisms may be available for **primary succession**. Such areas could be a garden pond filled with tap water, lava having erupted from a volcano, or even a concrete tile on a roof. The latter may become colonised by lichens. Occasionally an ecosystem may be destroyed, e.g. fire destroying a woodland. This allows **secondary succession** to begin, and signals the reintroduction of plant and animal species to the area.

Colonisation and succession also take place in water. Even an artificial garden pond would be colonised by organisms naturally. Aquatic algae would arrive on birds' feet.

- **Pioneer species (primary colonisers)** begin to exploit a 'new' habitat. Mosses may successfully grow on newly exposed heathland soil. These are the **primary colonisers** which have adaptations to this environment. Fast germination of spores and the ability to grow in waterlogged and acid conditions aid rapid colonisation. These plants may support a specific food web. In time, as organic matter drops from these herbaceous colonisers it is decomposed, nutrients are added to the soil and acidity increases. In time, the changes caused by the primary colonisers make the habitat unsuitable.

- Conditions unsuitable for primary colonisers may be ideal for other organisms. In heathland, mosses are replaced by heathers which can thrive in acid and xerophytic (desiccating) conditions. This is **succession**, where one community of organisms is replaced by another. In this example, the secondary colonisers have replaced the primary colonisers; this is known as **seral stage 1** in the succession process. Again, a different food web is supported by the secondary colonisers.

- At every seral stage, there are changes in the environment. The **second seral stage** takes place as the tertiary colonisers replace the previous organisms. In heathland, the new conditions would favour shrubs such as gorse and bilberry plus associated animals.

- The shrubs are replaced in time with birch woodland, the **third seral stage**. Eventually, acidic build up leads to the destruction of the dominant plant species.

- Finally, conditions become suitable for a dominant plant species, the oak. Tree saplings quickly become established. Beneath the oak trees, grasses, ferns, holly and bluebells can grow as a balanced community. This final stage is **stable** and can continue for hundreds of years. This is the **climax community**. Associated animals survive and thrive alongside these plant resources. Insects such as gall wasps exploit the oak and dormice eat the wasp larvae.

Jays are birds which eat some acorns but spread others which they store and forget. The acorns germinate; the woodland spreads.

Climax community

In Britain, an excellent example of a climax community is Sherwood Forest where the 'Major Oak' has stood for 1000 years. Agricultural areas grow crops efficiently by **deflecting succession**. Plants and animals in their natural habitat are 'more than a match' for domesticated crops. Herbicides and pesticides are used to stop the invaders!

Sometimes, conservationists have to prevent succession from occurring in order to protect habitats and species. The chalk grassland of the South Downs is an example of this. It contains some rare plant species that may disappear if deciduous woodland was to return.

Sample question and model answer

When given a passage, line numbers are often referred to. Try to understand the words in context. Do not rush in with a pre-conceived idea.

Read the passage, then answer the questions below.

line 1   Around the UK coast there are two species of barnacle, *Chthamalus stellatus* and *Balanus balanoides*. Both species are sessile, living on rocky sea shores.

The adult barnacles do not move from place to place but do reproduce
line 5   sexually. They use external fertilisation. Larvae resemble tiny crabs and are able to swim. At a later stage these larvae come to rest on a rock where they become fixed for the remainder of their lives.

The barnacles are only able to feed while submerged.

Adult *Chthamalus* are found higher on the rocks than *Balanus* in the adult
line 10   form as shown in the diagram below. Scientists have shown that the larvae of each species are found at all levels.

(a) Name the genus for each barnacle. [1]

   Chthamalus and Balanus

(b) What does the term sessile mean? (line 2) [1]

   is not motile, i.e. does not move from place to place

In this question are key terms, of which you will need to recall the meaning. This is only possible with effective revision. Did you already know the key terms **genus**, **sessile** and **motile**? Knowledge of these terms would enable you to access other marks, only easy when you have key word understanding. Try writing out a glossary of terms to help your long-term memory.

(c) Suggest how it is possible for neighbouring *Balanus* individuals to breed sexually (line 5) with each other even though they are sessile. [1]

   produce sperms which swim through the water

(d) Explain **one** advantage of the larvae being motile. [2]

   able to colonise new areas

   where there may be more nutrients/food

(e) Which type of competition exists between *Chthamalus* and *Balanus*? [1]

   interspecific competition

(f) Suggest an explanation for the distribution of each species of barnacle. [6]

   The larvae are found at all tide levels;

   at a lower tide level the barnacles are submerged for longer;

   Balanus may grow at a faster rate; can compete for food better; Chthamalus, which die out at lower levels

   near higher tide level the barnacles are exposed to open air for longer; Balanus may not be adapted to withstand desiccation; whereas Chthamalus can withstand drying out; and so survive without the competition from Balanus.

Practice examination questions

1 Ecologists wished to estimate the population of a species of small mammal in a nature reserve.

- They placed humane traps throughout the reserve and made their first trapping on day one, capturing 16 shrews.
- They were tagged then released.
- After day four a second trapping was carried out, capturing 12 shrews.
- Five of these shrews were seen to be tagged.

(a) The ecologists must be satisfied of a number of factors before using the 'capture, mark, release, recapture' method. List three of these factors. [3]

(b) Use the data to estimate the shrew population.
Show your working. [2]

(c) Comment on the *level* of reliability of your answer. [1]

[Total: 6]

2 The diagrams show stages in the development of a garden pond over a 10-year period.

butyl liner

A hole was dug, lined with butyl liner and new plants were placed in the pond.	Marginal plants grow, spread and die down in the winter. As they rot sediment falls to the bottom of the pond.	After a number of years the pond has completely covered over.
1990	**1995**	**2000**

(a) In 1990 irises, oxygenating pondweed and a water lily were planted in the pond. Algae were not planted but arrived in the pond in some other way.

(i) What term describes an organism that grows in a new habitat that previously supported no life? [1]

(ii) After a time the algae produced a thick 'carpet' of growth on the surface of the pond. Explain the effect this may have on organisms under the water. [5]

(b) Describe the stages which took place to produce the stable grassland after 10 years. [4]

[Total: 10]

Chapter 8

Energy and ecosystems

The following topics are covered in this chapter:

- *Energy flow through ecosystems*
- *Nutrient cycles*
- *Energy transfer and agriculture*
- *Effects of human activities on the environment*

8.1 Energy flow through ecosystems

After studying this section you should be able to:

- understand the roles of producers, consumers and decomposers in food chains
- understand the flow of energy through an ecosystem

LEARNING SUMMARY

Food chains and energy flow

AQA   4.5

Before energy is available to organisms in an ecosystem, photosynthesis must take place. Sunlight energy enters the ecosystem and some is available for photosynthesis. Not all light energy reaches photosynthetic tissues. Some totally misses plants and may be absorbed or reflected by items such as water, rock or soil. Some light energy which does reach plants may be reflected by the waxy cuticle or even miss chloroplasts completely! The energy that is trapped by photosynthesis and converted into biomass is called the **gross primary productivity (GPP)**.

> Around 4% of light entering an ecosystem is actually used in photosynthesis.
>
> KEY POINT

The green plant uses the **carbohydrate** as a first stage substance and goes on to make **proteins** and **lipids**. Plants are a rich source of nutrients, available to the herbivores which eat the plants. Some energy is not available to the herbivores because green plants **respire** (releasing energy).

The energy that is available to herbivores is called the **net primary productivity (NPP)**. It is calculated as follows:

Net primary productivity = gross primary productivity – energy lost in respiration

Energy is also lost from the food chain as **not all parts** of plants may be **consumed**, e.g. roots.

Food chains and webs

Energy is passed along a food chain. Each food chain always begins with an **autotrophic** organism (producer), then energy is passed to a primary consumer, then a secondary consumer, then a tertiary consumer and so on.

direction of energy flow →

Producer → primary consumer → secondary consumer → tertiary consumer
(herbivore)   (1st carnivore)   (2nd carnivore)

Note that a small bird is a secondary consumer when it eats apple codling moths but a tertiary consumer when it eats greenfly.

The producers always have more energy than the primary consumers, the primary consumers more than the secondary consumers and so on, up the food web. Energy is released by each organism as it respires. Some energy fails to reach the next organism because not all parts may be eaten.

Examiner's tip! Note that the primary consumer is at trophic level 2. It is easy to make a mistake with this concept. Many candidates do!

The following example shows three food chains linked to form a food web.

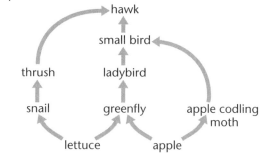

Each feeding level along a food chain can also be represented by a **trophic level**. The food chain below is taken from the food web above and illustrates trophic levels. Energy may be used by an organism in a number of different ways:

- respiration releases energy for movement or maintenance of body temperature, etc.
- production of new cells in growth and repair
- production of eggs
- released trapped in excretory products.

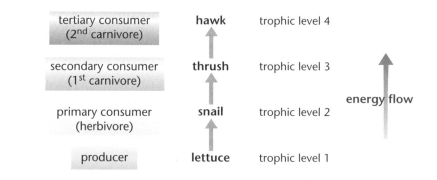

tertiary consumer (2nd carnivore)	**hawk**	trophic level 4
secondary consumer (1st carnivore)	**thrush**	trophic level 3
primary consumer (herbivore)	**snail**	trophic level 2
producer	**lettuce**	trophic level 1

energy flow

Pyramids of numbers, energy and biomass

A food chain gives limited information about feeding relationships in an area. Actual proportions of organisms in an area give more useful data. Consider a food chain from a wheat field. The pyramid of numbers sometimes does not give a suitable shape. In the example shown below, there are more aphids in the field than wheat plants. This gives the shape shown below (not a pyramid in shape!). A pyramid of biomass is more likely to be a pyramid in shape because it takes into account the size of the organism. It does not always take into account the rate of growth and so only a pyramid of energy is always the correct shape.

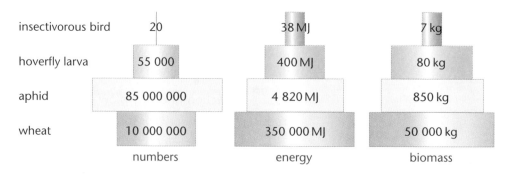

	numbers	energy	biomass
insectivorous bird	20	38 MJ	7 kg
hoverfly larva	55 000	400 MJ	80 kg
aphid	85 000 000	4 820 MJ	850 kg
wheat	10 000 000	350 000 MJ	50 000 kg

Biomass is the mass of organisms present at each stage of the food chain. The biomass of wheat would include leaves, roots and seeds. (All parts of the plant are included in this measurement.)

The organisms in the above food chain may die rather than be consumed. When this happens, the decomposers use extra-cellular enzymes to break down any organic debris in the environment. Corpses, faeces and parts that are not consumed are all available for decay.

8.2 Energy transfer and agriculture

After studying this section you should be able to:

- *understand a range of agricultural methods used to increase yield*

LEARNING SUMMARY

How are high yields achieved in agriculture?

AQA ▶ 4.5

The aim in agriculture is often to obtain good quality produce at maximum yield. Farmers grow crops and rear domesticated animals, like cattle, for meat and milk. Humans usually end the food chain as the top consumer. The human population is increasing constantly so **efficient** methods of agriculture have been developed.

Reduction of competition

Weeds reduce water, minerals and light reaching the crop plant; ultimately its rate of growth would be limited by the competing weeds. This is **interspecific competition** and takes place when different species need the same resources. Weeds can be removed **chemically** by use of a **herbicide** (weedkiller) or physically by an implement such as a rotavator which cuts up the weeds into tiny pieces, eventually killing them.

Intraspecific competition can also take place. This is where neighbouring plants of the same species compete for identical resources. This problem is reduced by making sure that crop plants are a suitable distance apart to achieve a maximum yield.

Use of fertilisers

It is important that crop plants have access to all the minerals they require to give a maximum yield. Farmers supply these minerals in fertilisers, usually in the form NPK (nitrogen, phosphates and potassium). By supplying them with these minerals, nitrogen is available to make protein, a key substance for growth. Phosphates help the production of DNA, RNA and ATP. Potassium helps with protein synthesis and chlorophyll production. Other minerals are also needed like iron and calcium. The more a plant grows, the more its biomass increases and usually the greater its surface area for light absorption. The amount of photosynthesis increases proportionally. If a farmer is to reach the maximum productivity of a crop, fertiliser is vital.

Increasing photosynthetic rate

As well as fertilisers helping to achieve a high productivity, other factors have a positive influence on growth:

- **irrigation** ensures that a plant has enough water for photosynthesis
- **suitable temperatures** can be achieved by use of a greenhouse to give ideal conditions for the process. If a gas heater is used then the **high concentration of carbon dioxide** excreted can be harnessed in photosynthesis.

Pest control

If pests such as aphids or caterpillars begin to damage crops then both quality and yield are reduced. Farmers combat pests in many different ways. A combination of different methods is called integrated pest management. This may include:

Examination questions on this topic often test knowledge of the advantages and disadvantages of each method. Chemicals may pass along food chains, and accumulate in greater quantities higher in the food chain due to the consumption of many smaller organisms, each carrying a small amount of the toxin. Biological control does not usually rely on a chemical agent so that the chemical risk is removed.

Biological control is usually much cheaper in the long term. If predators are used they go on to breed. Several repeat sprays of insecticides are needed through the growing season.

- **Pesticides**, sprayed onto crops, to kill pests. Chemicals used to kill insects are insecticides. **Contact insecticides** kill insects directly but **systemic insecticides** are absorbed into the cell sap. Any insect consuming part of the plant or sucking the sap then dies.
- **Biological control** which includes a range of different methods to get rid of pests.

  (a) The most commonly used method is to use **predators** to reduce pest numbers, e.g. in greenhouses infested with whitefly (*Trialeurodes vaporariorum*) the predatory wasp (*Encarsia formosa*) is introduced. The female wasp lays eggs into the scale (larva) of the whitefly. A young wasp emerges from each larva, having used the larva as a nutrient supply. The whitefly young are killed so its population decreases. Wasps increase in numbers and remain as long as some whitefly still remain.

  (b) **Pheromones** are also used. These are compounds secreted by organisms which affect the **behaviour** within the species, e.g. the apple codling moth larva spoils the fruit by tunnelling through apples. Female adult moths secrete a powerful chemical which attracts many males. This pheromone is now used in a trap. The sticky tent-shaped trap (below) shows how male moths are attracted and stick to the sides of the trap. Here they die and thousands of female moths out in the orchards are not mated and their eggs are not fertilised.

a codling moth trap

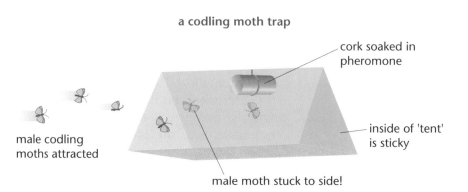

cork soaked in pheromone

male codling moths attracted

inside of 'tent' is sticky

male moth stuck to side!

  (c) **Irradiation** is used on insect pests which mate only once, e.g. the New World screw-worm fly lays eggs in cattle (and humans!). Larvae attack the internal systems having a devastating effect. Millions of screw-worm flies are bred then **irradiated**. They are subsequently released into cattle-producing regions. Irradiated males cannot produce fertile gametes. Any male mating with a female from the cattle fields results in unfertilised eggs so the population decreases. Cattle productivity is maintained.

- **Genetic engineering** can be used. A gene has been transferred to potato plants which enables them to produce a natural insecticide. This destroys 50% of aphids that attack the plants. The amount of damage is decreased.
- **Crop rotation** can be used to prevent build up of pests in the soil.

8.3 Nutrient cycles

After studying this section you should be able to:

- recall how carbon and nitrogen are recycled

The nitrogen cycle

AQA   4.6

Nitrogen is found in every amino acid, protein, DNA and RNA. It is an essential element! Most organisms are unable to use atmospheric nitrogen directly so the nitrogen cycle is very important.

There are three parts of the nitrogen cycle which are regularly examined:
- nitrogen fixation in leguminous plants
- nitrification
- denitrification.

nitrogen gas (N_2) in the atmosphere

Denitrification   lightning   Nitrogen fixation

(*Pseudomonas* bacteria)

free living nitrogen fixing bacteria in soil (*Azotobacter*)

In legumes (pea and bean family) *Rhizobium* **bacteria** live in nodules in the roots and produce **NH_4^+ ions** which enables the plant to make amino acids then **proteins**. Some carbohydrates are used by the bacteria.

other green plants

animal protein ← eaten by animals ←

death   faeces   urine

Decomposition by saprobiotic bacteria and fungi

artificial fertiliser

NO_3^- (nitrate) ← *Nitrobacter* bacteria ← NO_2^- (nitrite) ← *Nitrosomonas* bacteria ← NH_4^+ (ammonia)

← Nitrification →

Some important points

The association of *Rhizobium* bacteria with legume plants give advantages to both organisms. This relationship is known as **mutualism**.

Saprobiotic bacteria and fungi secrete extra-cellular enzymes.

The biochemical route from ammonia to nitrate is **nitrification**. This is helped by ploughing which allows air into the soil. Nitrifying bacteria are aerobic. Draining also helps.

- **Nitrogen gas** from the atmosphere is used by *Rhizobium* **bacteria**. These bacteria, living in nodules of legume plants, convert nitrogen gas into **ammonia (NH_3)** then into amine ($-NH_2$) compounds. The plants transport the amines from the nodules and make amino acids then proteins. *Rhizobium* bacteria gain carbohydrates from the plant, therefore each organism benefits.
- Plants support food webs, throughout which excretion, production of faeces and death take place. These resources are of considerable benefit to the ecosystem, but first **decomposition** by **saprobiotic** bacteria takes place, a waste product of this process is **ammonia**.
- Ammonia is needed by *Nitrosomonas* bacteria for a special type of nutrition (chemo-autotrophic). As a result another waste product, **nitrite (NO_2)** is formed.
- Nitrite is needed by *Nitrobacter* bacteria, again for chemo-autotrophic nutrition. The waste product from this process is **nitrate**, vital for plant growth. Plants absorb large quantities of nitrates via their roots.
- Nitrogen gas is returned to the atmosphere by **denitrifying bacteria** such as *Pseudomonas*. Some nitrate is converted back to nitrogen gas by these bacteria. The cycle is complete!

The carbon cycle

AQA ▶ 4.6

Carbon is the key element in all organisms. The source of this carbon is atmospheric carbon dioxide which proportionally is 0.03% of the volume of the air. Most organisms cannot use carbon dioxide directly so the carbon cycle is very important.

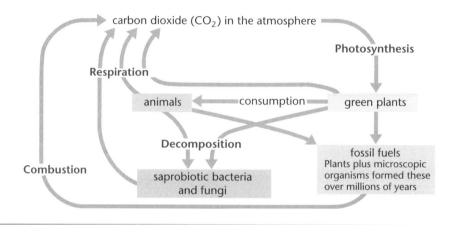

Some important points

- Producers carry out photosynthesis. This process incorporates the carbon dioxide into carbohydrates. These chemicals are used as a starting point to make lipids and proteins. Some of the carbon helps to form structures in the producers and some is released as carbon dioxide as a waste product of respiration.
- Producers are the starting point of food chains. After the plants are eaten by primary consumers carbon can be passed along to subsequent consumers. It can be incorporated into tissues, respired or excreted.
- Even after the death of a plant or animal, carbon dioxide can still be released. Saprobiotic bacteria and fungi respire using the organic chemicals in dead organisms as well as faeces and urine, etc.
- Compression of organisms millions of years ago resulted in the formation of fossil fuels. Combustion of these fuels releases carbon dioxide back into the atmosphere.
- The return of carbon dioxide to the air completes the cycle!

Other elements are also recycled. The decomposers have a major role in maintaining the availability of vital chemicals.

> Examiners often give a question about what happens to the energy in the chemicals of dead organisms or organic waste. Many candidates correctly state that microorganisms rot down the materials but then go on to state that energy goes into the ground. Big mistake! Energy is released by the respiration of the decomposers to support their life.

8.4 Effects of human activities on the environment

After studying this section you should be able to:

- *understand some causes and effects of pollution*
- *understand the effects of deforestation*

LEARNING SUMMARY

How human activities affect the environment

AQA ▶ 4.6

Activities carried out by the human population to supply food, power and industrial needs have considerable effects on the environment. These effects include atmospheric and water pollution, and destroying habitats and communities.

What is the greenhouse effect?

This is caused by specific gases which form a thin layer around the atmosphere. These gases include water vapour, carbon dioxide, methane, ozone, nitrogen oxides and CFCs. CFCs (CCl_2F_2, CCl_3F) have a greenhouse factor of 25 000 based on the same amount of carbon dioxide at a factor of 1.0. The **quantity** of the greenhouse factor gas needs to be considered to work out the overall greenhouse effect, e.g.

> carbon dioxide is 0.035% of the troposphere × greenhouse factor value
> 1 = 0.035
> CCl_2F_2 is 4.8×10^{-8}% of the troposphere × greenhouse factor value
> 25 000 = 0.012
> Water vapour is 1% of the troposphere × greenhouse factor
> 0.1 = 0.1

KEY POINT

It is clear that water vapour has the greatest overall greenhouse effect!

- The greenhouse gases allow short wavelength radiation from the sun to reach the Earth's surface.
- Some of the infra-red radiation fails to pass back through the greenhouse layer resulting in **global warming**.
- Polar ice caps may melt causing the sea to rise and subsequent reduction of land mass. Some aquatic populations could increase and some terrestrial populations decrease.
- Climatic changes are expected, so rainfall changes and heat increases will have significant effects.

Do not mix up the greenhouse effect with the 'hole in the ozone layer' – that is different! The ozone layer around the Earth absorbs some ultra-violet radiation from the sun. If a lot of ultra-violet radiation reaches the Earth's surface then many people succumb to skin cancer. CFCs cause a hole to form in the ozone layer. Not using these chemicals is the answer to this problem.

greenhouse effect

Follow the numbers to remember sequence in correct order!

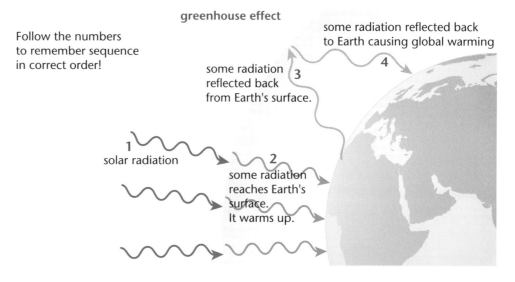

Deforestation

In countries such as Brazil, forests have been burned down. Large quantities of carbon dioxide and water vapour released into the atmosphere contribute to the greenhouse effect. However, the long-term effects are highly significant. Habitats and complete food webs are lost. Biodiversity is decreased so that many fewer species are represented on the land left after deforestation. The canopy of a forest intercepts and holds rain water, so too much rain does not reach the ground and cause flooding. Instead, much water evaporates back into the atmosphere. Without the forest, flooding is a danger, and without the tree roots, soil erosion takes place. If the deforestation was to make way for agriculture then there are major problems. Top soil is lost and nutrients leach into the ground and in the long term agriculture fails.

Water pollution

'Run off', containing fertilisers, enters rivers from fields. Similarly, sewage also pollutes rivers. The diagrams below show a river before and after sewage entry.

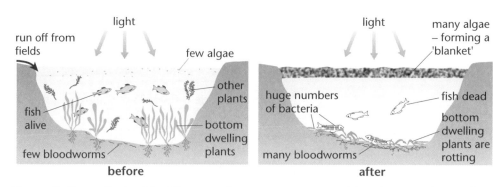

The polluting effect of fertilisers and sewage are caused by the constituent ions such as nitrates and phosphates. They result in **eutrophication**.

- **Nitrates and phosphates** enter the river and are absorbed by plants. This promotes plant growth.
- **Algae** float near the water surface and their population increases dramatically. A 'blanket' of algae soon covers the surface.
- Bottom dwelling plants do not share the same advantage. Initially the ions promote growth but **surface algae block the sunlight**. Plants beneath the algae die.
- **Bacteria** and other decomposers begin to **break down** the **dead plants** and some short lived algae. The **bacterial population increases** and proportionally takes more oxygen from the river water.
- **Fish die** as the oxygen content becomes much too low. Rotting dead fish contribute to even lower oxygen levels; again aerobic bacteria are responsible.
- Often there is an **increase in bloodworms** (tubifex). These are mud dwellers and possess a protein similar to haemoglobin which helps them to take in enough oxygen for survival, even at low concentration. Without fish to eat them numbers of bloodworms increase even more.

The sequence of stages of eutrophication are important. Take care to learn the correct sequence. Do not miss out any stage. Some candidates merely state that nitrates enter the river and fish die. Give more detail and score more marks!

How can the water pollution be measured?

There are many ways to measure both pollutants and their effects. Populations of algae, bloodworms or fish can be estimated. A key measurement is **biological oxygen demand (BOD)**. This is the amount of oxygen taken up by a sample of water at 20°C over 5 days. Clean water takes up much less oxygen than that polluted with organic material. Aerobic bacteria take up a large proportion of this oxygen. A river, heavily polluted with organic matter, has a very high BOD.

Indicator species

The presence or absence of a species in the river can be used as a sign of pollution. Mayfly larvae can only tolerate well oxygenated water. Bloodworms are only found in large numbers in water heavily polluted with organic matter. As the river flows downstream, organisms change the organic matter and eventually the oxygen content increases. A large population of mayfly larvae found in water downstream suggests that there is:

- a low BOD
- organic material further upstream which has been changed by bacteria so the water is no longer polluted.

Sample question and model answer

Some questions give information that you can use to deduce the answers (a) (i) is one of them!

(a) The sequence below shows how nitrate can be produced from a supply of oak leaves.

$$
\text{dead oak leaves} \xrightarrow{\text{decomposers}} \underset{\text{(ammonia)}}{NH_3} \xrightarrow[\text{bacteria}]{\textit{Nitrosomonas}} \underset{\text{(nitrite)}}{NO_2} \xrightarrow[\text{bacteria}]{\textit{Nitrobacter}} \underset{\text{(nitrate)}}{NO_3}
$$

This question targets the nitrogen cycle. Be ready to answer questions about *any part* of the cycle. Pure recall of the cycle is not enough! You need to apply your knowledge.

(i) Suggest the consequences of death of the *Nitrosomonas* bacteria. [4]

build up of ammonia; build up of dead leaves; death of Nitrobacter bacteria; no nitrite/no nitrate

(ii) Name the process by which bacteria produce nitrate from ammonia. [1]

nitrification

(iii) Name **two** populations of organisms not shown in the sequence which would be harmed by a lack of nitrate. [2]

denitrifying bacteria or Pseudomonas; plants or producers

(iv) Which organisms fix atmospheric nitrogen on the nodules of bean plants? [1]

(Rhizobium) bacteria

(b) Apart from adding fertiliser or irrigating a crop, how can a farmer make sure of producing a high yield ? [4]

In questions targeting yield in agriculture make sure that you have clarified all terms, e.g. do not confuse herbicide with pesticide.

make sure that the plants are the correct distance apart; use of pesticide or insecticide; use of herbicide or fungicide; use of biological control or named biological control; use of variety produced by selective breeding [any 4]

(c) (i) A farmer rears pigs by a factory-farming method. Pigs are kept indoors 24 hours per day in warm, confined cubicles.

How can this method result in the production of a greater yield of pork than from animals reared outside? [3]

Note that the reverse explanation for pigs reared outside can be given, e.g. more energy used for movement.

less energy is released for movement; less energy is used to maintain body temperature; more energy is used for biomass

(ii) Suggest why many consumers object to this factory-farming method. [1]

cruel or not ethical

(iii) Pigs are often given copper with their food because it promotes their growth. Suggest **one** disadvantage of using this method. [2]

it may contaminate the pork; people eat the pork and may be harmed

Practice examination questions

1 The number of species of grass and the number of leguminous plants growing in two fields was measured over a 10-year period. Field A was given nitrogenous fertiliser each year, but field B was given none. The results are shown in the graphs.

(a) (i) Suggest why there were fewer leguminous plant species in field A. [2]

  (ii) Suggest why there were more leguminous plant species in field B. [2]

(b) After the main investigation no fertiliser at all was used in either field. Cattle were allowed to graze in both fields. At the end of five years the number of legume species in each field had decreased. Suggest why the number of legume plants decreased. [1]

[Total: 5]

2 The following two gases help cause the greenhouse effect.

	Greenhouse effect factor	Relative amount in troposphere
Water vapour	0.1	1%
CFCs	25 000	4.8×10^{-8}

(a) Work out which gas has the greatest influence on the greenhouse effect. [2]

(b) Suggest **one** reason for the greenhouse effect resulting in:

  (i) an increase in the population of a species [1]

  (ii) a decrease in the population of a species. [1]

[Total: 4]

Chapter 9

Synoptic assessment

What is synoptic assessment?

You must know the answer to this question if you are to be fully prepared for your A2 examinations!

Synoptic assessment:

- involves the drawing together of knowledge, understanding and skills learned in different parts of the AS/A2 Biology courses
- requires that candidates apply their knowledge of a number of areas of the course to a variety of contexts
- is tested at the end of the A2 course both by assessment of investigative/ practical skills and by examinations
- is valued at 20% of the marks of the course total.

AQA assesses synoptic assessment in the A2 units by two types of question: structured questions carrying 15 marks and an essay question carrying 25 marks.

There is a structured question on Unit 4 requiring candidates to use subject matter and skills from Unit 4 and the AS specification. There is a structured question in Unit 5 requiring candidates to apply knowledge, understanding and skills from the AS and A2 specification. In the essay in Unit 5, candidates are required to use knowledge and understanding from across all units.

Practical investigations

You will need to apply knowledge and understanding of the concepts and principles, learned throughout the course, in the **planning**, **execution**, **analysis** and **evaluation** of each investigation.

How can I prepare for the synoptic questions?

- **Check out the modules** which will be examined for your specification's synoptic questions.
- Expect **new contexts** which draw together lots of different ideas.
- Get ready to **apply** your knowledge to a new situation; contexts change but the **principles remain the same**.
- In modular courses there is sometimes a tendency for candidates to learn for a module, achieve success, then forget the concepts. Do not allow this to happen! **Transfer concepts** from one lesson to another and from one module to another. Make those connections!
- Improve your powers of analysis – **take a range of different factors into consideration** when making conclusions; synoptic questions often involve both graphical data and comprehension passages.
- Less able candidates make limited conclusions; high ability candidates are able to consider several factors at the same time, then make a **number of sound conclusions** (not guesses!).
- You need to do **regular revision** throughout the course; this keeps the concepts 'hot' in your memory, 'simmering and distilling', ready to be **retrieved** and **applied** in the synoptic contexts.
- The bullet point style of this book will help a lot; back this up by summarising points yourself as you make notes.

Why are synoptic skills examined?

Once studying at a higher level or in employment, having a narrow view, or superficial knowledge of a problem, limits your ability to contribute. Having discrete knowledge is not sufficient. You need to have confidence in applying your skills and knowledge.

Synoptic favourites

The final modules, specified by AQA for synoptic assessment, include targeted synoptic questions. Concepts and principles from earlier modules will be tested together with those of the final modules. You can easily identify these questions, as they will be longer and span wide-ranging ideas.

> **Can we predict what may be regularly examined in synoptic questions?**
>
> 'Yes we can!' Below are the top five concepts. Look out for common processes which permeate through the other modules. An earlier module will include centrally important concepts which are important to your understanding of the rest.
>
> **KEY POINT**

Check out the synoptic charts

Synoptic links

Try this yourself! Think logically. Write down an important biological term such as 'cell division'. Link related words to it in a 'flow diagram' or 'mind map'. The links will become evident and could form the framework of a synoptic question.

1 Energy release

Both aerobic and anaerobic respiration release energy for many cell processes. Any process which harnesses this energy makes a link.

Examples

- Reabsorption of glucose involves active transport in the proximal tubule of a kidney nephron. If you are given a diagram of tubule cells which show both mitochondria and cell surface membrane with transporter proteins, then this is a cue that active transport will probably be required in your answer.
- Contraction of striated (skeletal) muscle requires energy input. This is another link with energy release by mitochondria and could be integrated into a synoptic question.
- The role of the molecule ATP as an energy carrier and its use in the liberation of energy in a range of cellular activity may be regularly linked into synoptic questions. The liberation of energy by ATP hydrolysis to fund the sodium pump action in the axon of a neurone.
- The maintenance of proton gradients by proton pumps is driven by electron energy. Any process involving a proton pump can be integrated into a synoptic question.

Energy: input and output

This has to be a favourite for many synoptic questions. Energy is involved in so many processes that the frequency of examination will be high.

2 Energy capture

Photosynthesis is responsible for availability of most organic substances entering ecosystems. It is not surprising that examiners may explore knowledge of this process and your ability to apply it to ecological scenarios.

Examples

- Given the data of the interacting species in an ecosystem you may be given a short question about the mechanism of photosynthesis then have to follow the energy transfer routes through food webs.
- Often both photosynthesis and respiration are examined in a synoptic type question. There are similarities in both the thylakoid membranes in chloroplasts and cristae of mitochondria.
- Many graphs in ecologically based questions show the increase in herbivore numbers, followed by a corresponding carnivore increase. Missed off the graph, your knowledge of a photosynthetic flush which stimulates herbivore numbers may be expected.

3 The structure and role of DNA

It is important to know the structure of DNA because it is fundamentally important to the maintenance of life processes and the transfer of characteristics from one generation of a species to the next. DNA links into many environmentally and evolutionarily based questions.

- The ultimate source of variation is the mutation of DNA. Questions may involve the mechanism of a mutation in terms of DNA change and be followed by natural selection. This can lead to extinction or the formation of a new species. Clearly there are many potential synoptic variations.

- DNA molecules carry the genetic code by which proteins are produced in cells. This links into the production of important proteins. The structure of a protein into primary, secondary, tertiary and quaternary structure may be tested. All enzymes are proteins, so a range of enzymically based question components can be expected in synoptic questions.

- The human genome project is a high-profile project. The uses of this human gene 'atlas' will lead to many developments in the coming years. The reporting of developments, radiating from the human genome project, could be the basis of many comprehension type questions, spanning diverse areas of Biology. Save newspaper cuttings, search the internet and watch documentaries. Note links with genetic diseases, ethics, drugs, etc.

4 Structure and function of the cell surface membrane

There are a range of different mechanisms by which substances can cross the cell surface membrane. These include diffusion, facilitated diffusion, osmosis, active transport, exocytosis and pinocytosis. Additionally glycoproteins have a cell recognition function and some proteins are enzymic in function. Knowledge of these concepts and processes can be tested in cross-module questions.

- In an ecologically based question the increasing salinity of a rock pool in sunny conditions could be linked to water potential changes in an aquatic plant or animal. Inter-relationships of organisms within a related food web could follow, identifying such a question as synoptic.

- In cystic fibrosis a transmembrane regulator protein is defective. A mutant gene responsible for the condition codes for a protein with a missing amino acid. This can link to the correct functioning of the protein, the mechanism of the mutation and the functioning of the DNA.

5 Transport mechanisms

This theme may unify the following into a synoptic question, transport across membranes, transport mechanisms in animal and plant organs. Additionally, they may be linked to homeostatic processes.

- The route of a substance from production in a cell, through a vessel to the consequences of a tissue which receives the substance, could expand into a synoptic question. Homeostasis and negative feedback could well be linked into these ideas.

Sample questions and model answers

A short question which cuts across the course. It refers back to AS. Do not forget those modules! See the *Letts AS Guide* for additional advice and those concepts not found in this volume.

Question 1 *(a short structured question)*

The kangaroo rat (*Dipodomys deserti*) is a small mammal that lives in the Californian desert. It has specialised kidneys so that it can live in these hot conditions.

(a) Name the genus that contains the kangaroo rat. [1]

Dipodomys

(b) What is the biological naming system called that gives the kangaroo rat its scientific name? [1]

binomial system

(c) Why is it important for any mammal to keep its body temperature close to 37°C? [2]

For correct enzyme action. Too hot and the enzymes will denature and too cold and the reactions will slow down.

(d) During the day the kangaroo rat spends most of the time underground. Suggest why this is. [2]

To keep cool. Otherwise too much moisture would be lost in keeping the temperature constant.

(e) The desert community that contains the kangaroo rat is the final product of succession in California. What is the name of the final, stable community that is produced by succession? [1]

a climax community

[Total: 6]

Question 2 *(a longer, more open-ended question)*

Plants and animals both need to exchange gases with the environment. Describe how animals and plants are adapted for efficient gaseous exchange. [10]

(Quality of written communication assessed in this answer.)

- *examples of respiratory surfaces in animals:*
  gills/lungs;
  tracheoles in insects;
  surface of protoctists;
  stomata in plants
- *large surface area:*
  way(s) in which this is achieved e.g.
  many alveoli;
  surface area/volume ratio in protoctists;
  many gill filaments;
  large surface area of leaves;
  many mesophyll cells
- *maintenance of diffusion gradients*
  way(s) in which this is achieved
  rich blood supply;
  ventilation mechanisms;
  sub-stomatal airspaces;
  spongy mesophyll air spaces;
  use of carbon dioxide in mesophyll cells

Sample questions and model answers *(continued)*

- small diffusion pathway
  barriers one cell thick;
  specialised cells, e.g. squamous epithelium;
  thin cell walls of palisade cells

Note, there is one mark available for legible text with accurate spelling, punctuation and grammar.

[Total: 10]

Question 3 *(a longer question of higher mark tariff)*

Different concentrations of maltose were injected in the small intestine of a mouse. The amount of glucose appearing in the blood and the small intestine after 15 minutes were measured. The results are shown in the graph.

> Prepare yourself for this type of synoptic question. It cuts across a large part of the specification. Make the links with different ideas. This fact is very important; concepts from AS are needed.

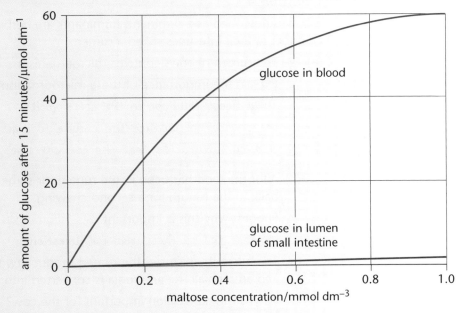

(a) (i) Describe the structure of a maltose molecule [2]

Two molecules of (alpha) glucose;

joined together by a glycosidic bond.

(ii) Maltose is converted into glucose by a hydrolysis reaction. What is a hydrolysis reaction? [1]

A reaction that breaks down a substance by the addition of water.

(b) Describe the effect of different maltose concentrations on the amount of glucose found in the lumen of the small intestine compared to the effect on the amount found in the blood. [2]

the maltose concentration has much more effect on the amount of glucose in the blood; the amount of glucose found in the blood is starting to level off but the amount in the lumen is increasing steadily.

> Even if you only cover one of these points, you can pick up a second mark by correctly using figures from the graph in your answer.

Sample questions and model answers *(continued)*

(c) The enzyme maltase is found on the cell surface membrane of the epithelial cells of the small intestine.

(i) How does the data on the graph indicate that the enzyme is not released into the lumen? [1]

Very little/no increase in the amount of glucose in the lumen.

This is a harder stretch and challenge question.

(ii) Explain why having the enzyme fixed to the cell surface will increase the rate of glucose absorption. [2]

Higher concentration of glucose produced close to intestinal lining;

will increase the concentration gradient between intestine and blood.

[Total: 8]

Question 4

A cow is described as a ruminant. Ruminants are herbivores that have a chamber in their intestines called a rumen.

(a) (i) The rumen of cows contains microorganisms.

Explain the importance of these microorganisms to the cow. [3]

They digest cellulose in the cow's food;

the cow cannot produce the enzyme to digest cellulose;

produce fatty acids that the cow can use.

(ii) After the food has been in the rumen for some time it is regurgitated back to the mouth for a second chewing.

Suggest why this is important. [1]

Increase the surface area for digestion.

(iii) The microorganisms in the rumen produce two waste products, methane and ammonia. The ammonia is converted into urea by the cow's liver.

Why is this conversion important for the cow? [1]

Ammonia is more toxic than urea.

(b) The table shows the amount of methane produced by different domesticated animals

Animal type	Methane production per animal in kg per animal per year	Total methane production in tonnes per year
buffaloes	50	6.2
camels	58	1.0
goats	5	2.4
sheep	6	3.4

(i) Which of the animals in the table are ruminants?
Explain how you can tell this. [2]

Buffaloes and camels;

they produce much more methane per animal.

Sample questions and model answers (continued)

This is a typical synoptic question as it links two different topics, digestion in herbivores and the greenhouse effect!

(ii) Which type of animal in the table is domesticated in the highest numbers?
Explain how you worked out your answer. [2]

Sheep; dividing the total methane production by the production per animal gives the highest number.

(iii) Methane is a potent greenhouse gas.
What is a greenhouse gas? [2]

A gas that prevents the escape of infra red radiation from the atmosphere; therefore causes the atmosphere to warm.

(iv) It has recently been discovered that methane is released when arctic ice melts.
Explain why people are concerned by this discovery. [2]

The release of methane would increase global warming;

which in turn would result in the release of even more methane.

[Total: 13]

Practice examination answers

Chapter 1 Energy for life

1

(a) in cytoplasm [1]

(b) pyruvate [1]

(c) 2 ATPs begin the process;
2ATPs are produced from each of the two GP molecules, so −2 + 4 = +2 ATPs net [1]

(d) animal; animal cells produce lactate [1]

(e) oxygen or aerobic [1]

[Total: 5]

2

(a) At this point the amount of carbon dioxide given off by the plant in *respiration*, is totally used by the plant in *photosynthesis*. [2]

(b) compensation point [1]

(c) The continued graph line falls (as light dims); line ends below the horizontal axis (when it's dark!). [2]

[Total: 5]

3

(a) mitochondrion [1]

(b) NADH [1]

(c) cytochrome [1]

(d) ATP [1]

[Total: 4]

4

(a) (i) rate of photosynthesis is proportional to light intensity; rate limited by amount of light available

(ii) as light intensity increases it results in significantly less increase in the rate of photosynthesis

(iii) rate of photosynthesis has levelled off, no longer limited by light (but other conditions could be limiting!). [3]

(b) Similar shape of graph, begins at origin, but graph line above the given plotted curve. [1]

[Total: 4]

Chapter 2 Response to stimuli

1

(a) (i) IAA (at these lower) concentrations is *proportional* to the angle of curvature of the stem. [1]

(ii) IAA (at these higher) concentrations is *inversely proportional* to the angle of curvature. [1]

(b) *More* IAA causes the cells at side of stem in contact with agar block to elongate more than other side.

So this side grows more strongly bending stem towards the weaker side. [2]

(c) Growth is only stimulated up to a certain high IAA concentration, after this curvature would be inhibited. [2]

[Total: 6]

2

(a) A = actin; B = myosin [2]

(b) action potential reaches sarcomere [1]

(c) both filaments slide alongside each other;
they form cross bridges;
during contraction the filaments slide together to form a shorter sarcomere [2]

[Total: 5]

3

(i) resting potential achieved; Na^+ / K^+ pump is on [2]

(ii) Na+ / K^+ pump is off; so Na^+ ions enter axon [2]

(iii) maximum depolarisation achieved; K^+ ions leave [2]

(iv) Na^+ ions leave due to Na^+/ K^+ pump being back on; this is during the refractory period;

(v) at end of this resting potential re-established; axon membrane re-polarised [4]

[Total: 10]

Chapter 3 Homeostasis

1 (a)

	Nervous system	Endocrine system
Usually have longer lasting effects		✓
Have cells which secrete transmitter molecules	✓	
Cells communicate by substances in the blood plasma		✓
Use chemicals which bind to receptor sites in cell surface proteins	✓	✓
Involve the use of Na^+ and K^+ pumps	✓	

[2]

(b) homeostasis [1]

[Total: 3]

2

(a) Moisture evaporates from the dog's mouth; takes heat away from the body; cannot sweat from the skin because of fur. [2]

(b) Vasoconstriction occurs; less blood flows close to the skin's surface. [2]

Chapter 3 Homeostasis (cont.)

(c) Sweat cannot evaporate so easily; as air is saturated with water vapour; therefore, less heat is lost from the skin. [2]
[Total: 6]

3

(a) **B**, because as glucose levels rose after meals they did not decrease enough (this kept the blood glucose level too high) [1]

(b) glucose levels fell after every meal, so glucose must have entered the cells and liver [1]

(c) in the pancreas;
in the β cells of islets of Langerhans (max 2) [2]
[Total: 4]

Chapter 4 Further genetics

1

(a) no immigration and no emigration; no mutations; no natural selection; true random mating; all genotypes must be equally fertile [4]

(b) (i) $q^2 = \dfrac{48}{160}$

$= 0.3$

$q = 0.55$

but $p + q = 1$
so $p = 1 - 0.55$
$= 0.45$
but $p^2 + 2pq + q^2 = 1$
so $0.45^2 + 2 \times 0.45 \times 0.55 + 0.55^2 = 1$
$0.2 + 0.5 + 0.3 = 1$
BB $= 0.2$ Bb $= 0.5$ bb $= 0.3$ [3]

(ii) BB 2000   Bb 5000   bb 3000 [2]
[Total: 9]

2

A (iv), B (iii), C (v), D, (ii), E (i). [Total: 5]

3

(a) triplet [1]

(b) codes for an amino acid, codes for stop or start [2]
[Total: 3]

Chapter 5 Variation and selection

1

(a) continuous variation; [1]

(b) two from:
genetic;
the nutrition of the mother;
mother's smoking;
mother's alcohol intake;
mother's health; [2]

(c) three from:
heavy babies have higher death rate;
light babies have higher death rate;
so babies of average mass more likely to survive;
they are more likely to have babies of average mass; [3]

(d) Modern techniques can increase survival of light and heavy babies; they in turn will reproduce; [2]
[Total: 8]

2

(a) **Allopatric speciation** takes place after geographical isolation;
• the rising of sea level splits a population of animals; formerly connected by land creating two islands;
• mutations take place so that two groups result in different species.

Sympatric speciation takes place through genetic variation;
• in the same geographical area;
• mutation may result in reproductive incompatibility;
• perhaps a structure in birds may lead to a different song being produced by the new variant;
• this may lead to the new variant being rejected from the mainstream group;
• breeding may be possible within its own group of variants. [6]

(b) Mate them both with a similar male, to give them a chance to produce fertile offspring.
• If they both produce offspring, take a male and female from the offspring, mate them,
• if they produce fertile offspring then original females **are** from the same species. [2]
[Total: 8]

Chapter 6 Biotechnology and genes

1

(a) Identify the specific section of DNA which contains the gene; this can be done using reverse transcriptase; insert DNA into a vector/insert into *Agrobacterium tumefaciens*; this bacterium/this vector then passes the DNA into the recipient cell. [5]

(b) herbicide kills weeds; which reduces competition; for light or water or minerals; soya plants unharmed [3]

[Total: 8]

Chapter 7 Ecology and populations

1

(a) no significant migration;
no significant births or deaths;
marking does not have an adverse effect. [3]

(b) S = total number of individuals in the total population
S_1 = number captured in sample one, marked and released, i.e. 16
S_2 = total number captured in sample two, i.e. 12
S_3 = total marked individuals captured in sample two, i.e. 5

$$\frac{S}{S_1} = \frac{S_2}{S_3} \quad \text{so, } S = \frac{S_1 \times S_2}{S_3}$$

$$S = \frac{16 \times 12}{5}$$ Estimated no. of shrews is 38 [2]

(c) Not very reliable because the numbers are quite low. High population numbers are more reliable. [1]

[Total: 6]

2

(a) (i) pioneer or primary coloniser [1]

(ii)
  • algae cut off light from plants underneath;
  • they die as a result;
  • bacteria or fungi or saprobiotics decay them;
  • they use a lot of oxygen;
  • fish die due to not enough oxygen;
  • blood worms increase in number as they are adapted to small amounts of oxygen.

[any 5 points] [5]

(b) • marginal plants or irises were introduced;
  • they spread;
  • each year the foliage died and rotted;
  • this organic material or humus added to the soil or mud;
  • secondary colonisers spread from other areas;
  • succession took place.

[any 4 points] [4]

[Total: 10]

Chapter 8 Energy and ecosystems

1

(a) (i) When given fertiliser the grasses competed for resources better that the legumes; some legume species could not grow in these conditions. [2]

(ii) Without fertiliser the grass species did not have enough minerals so did not compete as well; the legumes fixed nitrogen in root nodules so could grow effectively. [2]

(b) Cows grazed on some species more than others/ perhaps trampling by cattle destroyed some species but others were tougher and survived/perhaps waste encouraged the growth of some species whereas others were destroyed. [1]

[Total: 5]

2

(a) greenhouse effect factor x amount
= water vapour
CFCs 25 000 x 4.8 x 10^{-8} % = 0.012
water vapour has greatest greenhouse effect [2]

(b) (i) new sea areas so more marine organisms or named organism/formerly cold area grows new warm-climate plants [1]

(ii) deserts formed which reduce food availability/ cold-adapted organisms not suited to new climate/terrestrial organisms destroyed by the rising seas [1]

[Total: 4]

Index